WHEN WE CAME HOME

WHEN WE CAME HOME

HOW THE VIETNAM WAR CHANGED THOSE WHO SERVED

Jack McCabe

OddInt Media

Front cover photograph by John Veers, Vietnam veteran 1971-1972. Back cover photograph of the Wall by Tom McCabe, Coastal Light Photographic Arts.

Cover design by Jason Vargas and Lee Sievers.

Published by OddInt Media
First published 2017

ISBN 978-0-9797-8686-0 (paperback)
ISBN 978-0-9797-8689-1 (ebook)

This book is dedicated to the fallen, who shall forever be young, and to those who died later as a result of this war, a war that took their youth so many years ago.

Ambushed

I call out
My voice
Caught in the
Vortex of
Inner space

Asking why

Unending echoes
Ricochet across
My mind

As Vietnam
Once again
Ambushes
My soul

In Peace,
Penni Evans
Donut Dollie, 1970-1971

Table of Contents

I would like to thank the following who were instrumental in the writing of this book:

My wife, Pat, who corrected my grammar and helped me organize this work and put it in a workable format. She listened to me tell these stories countless times, boring our friends.

Jenny La Sala (author of *Comes a Soldier's Whisper*, *Vietnam & Beyond: Veteran Reflections*, *When Daddy Comes Home*, and *Never Forgotten: The Vietnam Veteran Fifty Years Later*), for her help in setting up my Facebook page and Twitter account, and for her constant encouragement and advice on this project and how to proceed when I seemed lost.

John Brennan (author of *Vietnam War Helicopter Art, Volumes 1 & 2* and *Vietnam War U.S. Army Helicopter Names, Volumes 1 & 2*), who referred Vietnam veterans to me for interviews.

Thank you to everyone who contributed to this book and allowed me to interview them. I may not have used your interview but you are all part of this book.

Preface

It was April 1975. I had been home from Vietnam almost three years. I was living at home with my parents in their small house on the northwest side of Chicago while attending school. I was a very angry young man but didn't realize it.

I was sitting in my parents' family room watching the news. The North Vietnamese Army was overrunning South Vietnam. Everything was falling apart. The South Vietnamese government was collapsing, and the NVA was taking over province after province. I watched my old base at Phu Loi fall, and then Saigon fell.

I remember watching the TV and talking back to it. I was angry, and swearing loudly. I was bitter and outraged at the outcome of the rapidly ending war. In between rants, I glanced into the kitchen and saw my mother standing at the sink, softly crying. I was shocked because my mom rarely cried. I didn't understand what was wrong, so I turned off the TV and walked over to her. I put my hand on her shoulder and asked her what was wrong. She just kept crying and through her tears said, "I just want to see you happy again."

I was dumbfounded. I didn't know what to say. I was happy…wasn't I? I quietly left the house and went for a walk around the block to sort things out. I was confused and felt horrible that my mother was crying. I couldn't imagine what I had done to make her so miserable. I didn't think there was anything wrong with me. I didn't understand. I felt completely normal but, obviously, I wasn't. Or, I guess I was as normal as I could be at that point in my life.

Thirty-eight years later, after attending my army unit's reunion at the Wall, I had two angry outbursts that totally surprised me. Both were directed at the one person I should never feel angry at—my wife. They were over some stupid, trivial things that should never matter but always seem to. I decided it was time to go to the VA and see a doctor. I met with a psychologist and the first thing he asked me was, "Why are you here?" I responded, "I have been pissed off for the last 40 years, and I don't want to go into the ground pissed off."

Introduction

HOME, HOME! That is all we thought about in Vietnam. Coming home to our family, our friends, our wives and our girlfriends. We processed out, turned in our rifles and all our gear. We caught a chopper to Bien Hoa, Cam Ranh Bay, or Da Nang and waited for our Freedom Bird to take us out of Hell. Some cheered as we left Vietnam airspace, some cried, and some were silent. In two short days, we would be home. From Hell to home in two days. The shock of re-entry into "The World" was similar to the shock of battle the first time.

Most us were met with protests, anger, and indifference. The military warned some of us not to wear our uniforms, as did some of the airlines. They felt the uniform would spark protests. We should have felt proud or at least lucky to have survived, but we didn't. We came home to a hostile world. Our true brothers and friends were back in Nam.

Our families didn't understand us. Our friends had moved on with their lives. The public didn't want anything to do with us. No one asked about Vietnam or what we, as individuals, did there. We hid that part of our lives. Over 58,000 of us died there. More than 350,000 more were wounded. There would be no parades, no free drinks, no "Welcome Home." We buried our resentment and distrust for many years, buried our feeling of abandonment. Buried it all in work, insomnia, depression, dejection, anger, alcohol and, for some, drugs.

Now we can see our scars. No one came home whole.

This book tells my coming-home story and the stories of many other Vietnam veterans—US Army, Navy, Air Force, Marines, Nurses, WACs, Red Cross Donut Dollies, and Australians. They look back and reflect on the war and how it changed them. It changed everyone who was there, no matter what job you held—cook, mechanic, grunt, pilot.

More than 150 veterans were interviewed for this book. Their stories were heartfelt and, for many, very emotional. Some decided later that they didn't want their stories published: they just needed to talk. The most important outcome for this book is to give Vietnam veterans a voice.

Many Vietnam veterans are now suffering from the effects of Agent Orange, which is slowly killing of them. It has caused birth defects in our children, and will take seven generations to purge. We have a VA whose bureaucracy baffles and frustrates us. Many veterans are still battling the effects of PTSD, and fighting to win legitimate claims.

Jack McCabe

Chapter 1 - Our War

I grew up on the northwest side of Chicago. A kind of crazy neighborhood with a fascinating cast of characters. It was a great childhood, full of happy and fun times. It was safe back then, and we would always be out playing in the streets until the streetlights came on, then we had to go home for dinner. After dinner, we would rejoin the neighborhood kids and play until our mothers would call us home by yelling our names from the front steps.

When we weren't playing sports, we were playing war in the streets, yards, alleys, and forest preserves. We fought imaginary battles against Japanese and German infantry and tanks. Of course, we won every battle! Americans always win!

We were just kids growing up in a typical Chicago neighborhood, happy and innocent. But there were things going on in the world that would have a lasting effect on us. The Cold War was in full swing. At my house, we watched the news every night while eating dinner and saw the tension build between the United States and Russia. Kennedy and Khrushchev were at strong odds, and it was feared there may be another war. A nuclear war!

Every Tuesday morning at 10:00 AM the air raid sirens would sound and at school we would have an air raid drill. We would all crawl under our desks and cover our heads with our hands until the sirens stopped. Our school had metal signs that proclaimed it a "Fallout Shelter." At O'Hare Airport there were four black ready hangers: in these hangers were four fighter jets, armed and ready to take off at a moment's notice to intercept Russian bombers. It was a scary time for a child.

Everyone in our neighborhood was familiar with war. Every grown man on the block had served in World War II. My dad had spent two years in the Pacific with the 13th Air Force. The man next door had been in Europe with the 8th Air Force. He was a bombardier, was shot down, and became a prisoner of war. One of my friend's fathers had been in the Navy—PT Boats in the Solomon Islands and the Philippines. Across the street, we had a Navy veteran who had served on a destroyer in the

Atlantic, and an 82nd Airborne veteran who made four combat jumps and had been wounded in Europe.

My father's story was probably typical of many G.I.s in World War II. He was drafted into the army in 1941 for one year, which was the norm back then. This was before the Japanese attacked Pearl Harbor. He was trained as an infantryman and then as a tanker, and was stationed at Camp Roberts, California, when the war broke out.

After the war began he was sent to Signal School. As the country mobilized after December 7, 1941, the influx of new troops required soldiers to train them. My dad was then assigned as a drill instructor. Eventually, he was appointed a warrant officer and, like so many others, was sent to the South Pacific. When he arrived at New Caledonia, he was assigned to the 13th Air Force, 319th Fighter Control Squadron, and sent to Bougainville, Solomon Islands. His job was to call in and direct fighter aircraft to intercept Japanese air attacks and to direct air strikes on Japanese troop concentrations.

The Marines and Army slowly took over Bougainville from the Japanese. At one point the Japanese massed for an attack on Torokina Airfield. My dad called in airstrikes from the fighters stationed there. They took off, dropped their bombs, landed, re-armed, and bombed again. The Japanese attack was destroyed before it began. It was estimated that over 600 Japanese soldiers were killed. Dad went down and looked at the carnage after it was all over. The sight of the destroyed humanity never left him. His experiences on Bougainville changed him forever. They haunted him when he had dementia. He would look at me with a wild look in his eyes and tell me, "We've got to check the lines! The Japs are coming!"

But, Bougainville was not the end of his war. It was only the beginning. He fought the Japanese for over two years, participating in additional campaigns in New Guinea and the Philippines. He was in the Philippines preparing for the invasion of Japan when the atomic bombs were dropped on Hiroshima and Nagasaki. With the surrender imminent, he and his troops loaded on an LST and sailed for Japan. He actually landed in Japan before the surrender was signed. The war was over.

After several months of occupation duty, his entire unit boarded a transport ship for the trip home. The trip across the Pacific took a few weeks, and during this slow voyage Dad and his fellow soldiers discussed what they would do when they got home. They talked about the things they did in the war and the things they saw. They tried to make sense of it all. It had been a long war, but now it was over, over for everyone. It was the end of 1945, and Dad had been in the Army Air Force for over four years.

Their ship pulled into San Francisco, sailing under the Golden Gate Bridge. Virtually everyone was top-side. He told me it was a crisp, crystal-clear day. As they approached the dock, they could see a crowd waiting for them. There was a band playing, everyone was cheering and waving, and there were confetti in the air. It took a long time to debark from the ship. Trucks were waiting for them and took them to Oakland Army Depot. They were told they would be given a steak dinner and a beer, compliments of the US Government. As they filed into the mess hall for their steak dinner, they discovered they were being served by German POWs—healthy, fat, German POWs who were very happy, greeting them with smiles. Here were the Americans, fresh from the Pacific: skinny, many of them sick from years of poor diet and tropical diseases, yellow from anti-malaria pills. Who, they asked each other, who won the war?

After processing and having received their campaign ribbons, they dressed in their Class A uniforms and my father and a friend from Chicago headed to San Francisco for a couple of drinks. They couldn't buy one. Everyone bought them drinks. And, they drank their fill. Dad said the mood in San Francisco was magical.

The next day, sporting hangovers, the two headed to the train station to try to catch a train bound for Chicago. There were thousands of soldiers, sailors, and Marines trying to get home. He had sent my mom a telegram telling her he was in the States but had no idea how long it would take to get home. Finally, he and his friend caught a train for

Chicago. It took three days to get there. The train pulled into Union Station, they grabbed their bags and walked out into the cold Chicago air. With Japanese rifles sticking out of their duffel bags they prepared to part company and head home. But neither one was in a hurry to leave. There was something scary about going home, something foreign and unknown. So they went to the Drake Hotel and had a couple of drinks.

Finally, they realized that the time had come to get on with the next phase of their lives. They swore they would keep in touch then went their separate ways. My father took a cab to his in-laws' house where my mom was living. He told me that during the cab ride home he felt like he was going into combat. He was stressed, nervous, and very apprehensive. He had been gone for over two years. Would he and my mom even know each other? Would they get along? My mom did not know exactly when he would be home as he hadn't called when he arrived in Chicago. The cab finally pulled up to the house on Roscoe Street. Home. Of course, everyone was extremely happy to see him, and slowly he and my mom started to put their lives back in order after such a long separation.

I thought of this story often over the years.

In June of 1964 I had just finished the seventh grade and was excited about starting eighth grade in the fall. I was growing up. The whole world was ahead of me. Life was filled to the brim. It was a perfect summer. The weather was typical for Chicago—beautiful summer days, occasional thunder storms and some 90+, high humidity days. But, in the distance storm clouds of a different type were gathering. We couldn't see them from Chicago but they were there, lurking, ready to pour. Our world was about to change.

On August 2, 1964, the USS Maddox was performing a signal intelligence patrol in international waters off the coast of North Vietnam. According to reports, three North Vietnamese Navy patrol boats approached and the Maddox engaged. During this battle, the Maddox fired over two hundred three-inch and five-inch shells at the North Vietnamese craft. US aircraft also strafed the boats. They were all damaged, but none was sunk. One of our aircraft was damaged, and a

14.5mm round hit the Maddox without wounding any US sailors. Reportedly, four North Vietnamese sailors were killed, and an additional six were wounded.

There was a reported second incident two days later, August 4. It was reported that North Vietnamese torpedo boats once again were attacking US naval vessels. Was this an act of war? In reality, this action may have involved "Tonkin Ghosts" (false radar images) and not actual North Vietnamese Navy torpedo boats. In fact, Secretary McNamara admitted in Errol Morris' 2003 documentary, "The Fog of War," that the second attack never occurred.

The Gulf of Tonkin Incident, as it was called, resulted in Congress passing the Gulf of Tonkin Resolution on August 7, 1964. This resolution gave President Johnson the authorization to take any measures he believed necessary to retaliate. President Johnson signed it into law on August 10. The stage was set.

In March 1965, the first US combat troops arrived in Vietnam. Some 3,500 US Marines landed in Da Nang, South Vietnam, joining the 23,000 American advisors already there. Throughout 1965, the American presence ballooned as more and more combat troops were sent into the fray. Support troops began to arrive by the thousands. The war was on.

Troops streamed into Vietnam at an extremely rapid rate. Whole units up to division size were shipped to Vietnam. It seemed that almost everyone in the service at that time wound up being sent there. Then the casualties began to come home to be laid to rest by grief-stricken families. As the number of casualties mounted so did the protests. The country was torn apart by the war. Hawks vs. Doves. Angry clashes between students and police. The returning veterans were caught in the middle, symbols of the military establishment.

At my parents' house we watched the war unfold every night on the evening news while we ate dinner. "Vietnam" became a household word. Initially the airports were full of uniformed servicemen. Towards the end more and more returning G.I.s were wearing civilian clothes. But, you could tell who they were. Short hair, tanned skin, and a faraway look in their eyes. The thousand-yard stare. These veterans didn't advertise their service but rather they kept quiet about it. They blended back into society and disappeared amongst the silent majority. Most people did not want to know about them or their experiences.

The war became very real to our family when my brother Tom dropped out of high school in 1966 and joined the Marine Corps. My dad was not happy about him dropping out, but was extremely proud of his enlistment. I am sure he was very apprehensive, as he had been to war and knew what could happen to his son now that Vietnam was heating up.

Although the Vietnam War did not hit our neighborhood very hard (very few neighborhood kids served in the military), June 1969 did turn out to be a wake-up call for us. Larry Davison, who lived around the corner, joined the Marine Corps in July 1968. He was trained as a rifleman and was sent to Vietnam on 20 May 1969. He was assigned to the 26th Marines, 3rd Marine Division in the field. Larry was killed in action on June 7, 1969, only 18 days after arriving in-country. His death cast a black cloud over the neighborhood. About this time my brother Tom was heading to Vietnam and I was approaching draft age.

Most of my classmates avoided the draft by attending various schools, including the local junior college, Triton. The often-quoted phrase in my high school was, "Triton's better than fight'n'."

In my family, college was not an option. You either went to work in a factory or joined the armed forces. I was 17 when I graduated high school so I went to work at a local factory, but the work did not suit me. I joined the Army two days after my 18th birthday, and less than a year later I was in Vietnam. My brother returned from Vietnam in June 1970, and I went over a few months later. I can only imagine how hard it was for my parents.

Unlike the WW II veterans, our transition back to the real world was jarring. Many were out in the field one day and three days later were back home in their living room. Unlike our fathers, who were proud to wear their uniforms home, we were asked by the airlines and warned by the military not to wear uniforms, to avoid protests and provocation. There were no parades, no bands, no free drinks, or welcomes of any kind. We came home individually in a plane full of strangers. When we arrived in the States we all went our separate ways. Many were met by a hostile public, and many more were met with indifference. Organizations such as the VFW and the American Legion, usually a source of brotherhood for the returning G.I., did not welcome us and in some cases scorned us. We were part of an unpopular war that we didn't "win" in their eyes. This is why the Vietnam Veterans of America was formed: its slogan is, "Never again will one generation of veterans abandon another."

Nonetheless, not all Vietnam veterans had a bad homecoming; some came home without encountering trouble, protests, or problems. But, like all war veterans, they were changed forever by their experiences in war. No matter what job they did, branch of service they were in, or where and when they served, they would never be the same. Although most experiences are universal, the Vietnam experience is unique in many ways, especially when compared to World War II.

On a personal level, I would have to say that I guess I never felt the same after coming home. I was angry and bitter. I was not a grunt. I did not beat the bush and engage in heavy combat. Yet when I came home I felt distant from my friends and family. My high school friends told me that they decided to give me a year to get back to normal. Normal? What was normal? How long should I give them? Things went downhill from there. I haven't seen any of them for many, many years, nor do I ever wish to.

As I talked to Vietnam veterans I came to find out that most of us felt the same way to one degree or another. We all had our problems adjusting. For us there was no closure. While in Vietnam we counted the days until we would come home, but when the time finally came it was anti-climactic. "Here are your orders. You can go home now." That's it? The war was still going on just like it was when we arrived. Nothing had changed. What had we accomplished? Were the buddies we were leaving behind going to be okay?

When the last US troops left Vietnam in 1973 there were no celebrations around the country as there had been at the end of World War II. It was just another day. For us, our war ended individually when our tour was up. It was different for every one of us who served. We packed our gear and went home. Our war was over. But for many, the war never really ended. It just stopped. People tell us to "put it behind you, forget about it, it was a long time ago," and to "get over it." We don't seem to be able to do that. Unless you have experienced Vietnam, you may never understand.

Why did I decide to write this? I feel this is a story that needs to be told. Many Vietnam veterans are still suffering from issues brought on by this war—PTSD, anger, substance abuse, withdrawal from society, broken relationships, Agent Orange illnesses, etc. I am not a psychiatrist, psychologist, or sociologist, but I know enough veterans to know that the war and the way we were treated have a lot to do with the problems we Vietnam veterans have today. Bob Greene, in his book, "Homecoming," addressed many of these issues. But that book was published in 1989. A lot of time has passed. The question now is, "How are we, as aging Vietnam veterans, doing today?" Are we still suffering the effects of that war? Have we moved on like so many people tell us? Or, are we still angry and hurt by the way our peers, the public, and our government treated us? I will let the veterans speak for themselves.

Chapter 2 - Army Stories

By 1965, there were approximately 23,000 army advisors and aviation personnel serving in the Republic of Vietnam. That summer, the First Cavalry Division and a brigade of the 101st Airborne Division were deployed to South Vietnam. Headquarters, US Army Vietnam (USARV), was established on July 20, 1965, and stationed in Long Binh, which grew to be the largest US base in Vietnam. Troop concentrations in South Vietnam swelled to over 180,000 troops by the end of 1965. And by the end of 1968, US troop strength increased to more than 536,000.

Approximately 1,407,000 soldiers served on the ground in South Vietnam: 38,224 Army personnel died there, 96,802 were wounded, requiring hospital care; and 104,723 were wounded that did not require hospital care.[1] No matter what job you had or where you were stationed, there were casualties in your Military Occupational Specialty. Vietnam proved to be hazardous to your health. There was no safe place in Vietnam.

As the war dragged on, the number of protests increased. Morale decreased and racial problems increased. The war changed. The Army changed. No one wanted to be the last soldier to die in Vietnam. Those of us who were lucky enough to come home couldn't wait to leave, but weren't welcomed back. It didn't matter if you were drafted or had volunteered. We were all treated the same: poorly. We veterans lived in limbo, a lost world between Vietnam and Home. Some still live there, in No Man's Land.

1. Casualty figures from Defense Casualty Analysis System (DCAS)

Staff Sergeant Loyd Cates
Co. D, 5/12, 199th Light Infantry Brigade

I was drafted into the Army in July of 1968 for two years. I attended basic training at Fort Polk, Louisiana, and then infantry training immediately afterward. I had been going to college but I dropped out for one semester. I needed money so I went to work in the oil fields. That's when I got drafted. I was only out of school for three months but that was long enough.

Right after AIT I was sent to NCO school at Fort Benning. I think the course was about nine weeks long. It was extra training and I felt I needed that as I was sure I was going to be sent to Vietnam. Actually, there was never any doubt in my mind that I would be sent to Vietnam, so it was no surprise when I received my orders.

I went to Vietnam in early July 1969. I was assigned to Company D, 5/12, 199th Light Infantry Brigade. When I arrived the 199th was based near Xuan Loc, northeast of Saigon. The had recently moved here from the Delta. From rice paddies to jungle. We actually worked out of fire base Libby.

I was wounded in October of 1969. We were ambushed and I was hit with mortar fragments. I was lucky. I had fragmentation wounds pretty much all over my body, but none of them was that bad. I think I spent maybe a week in the hospital. I still have some little pieces in me, those that were too small for them to address. For the first three or four years after I was back home some of those little pieces worked their way out.

In May or June 1970, the 199th was ordered to Cambodia to take part in the incursion there. I was offered a job back in rear because I had actually been in charge of my platoon for most of the time that I was there. And, of course, I jumped on that like a dog on a bone. It really turned out not to be much of a job at all, but I missed out on going to Cambodia. At this time the brigade was based at Long Binh and that's where I was when I received my orders to go home.

Coming Home

I flew home out of Bien Hoa in July 1970. We stopped in Tokyo, and, then, on to Oakland. When I arrived in Oakland I went through about 18 hours of processing. After I finished processing I was out—discharged. They offered us a steak dinner but I bypassed that so I could get to the airport quicker.

To be honest with you I can't remember the flight home. Why, I don't know. I don't know if I just slept the whole way or just blocked that out. I flew from San Francisco to Los Angeles, and then Los Angeles to Shreveport, Louisiana. I never saw any protesters the entire time.

When I arrived in Louisiana I called my home, which was about 60 miles from Shreveport. One of our neighbors answered the phone, which was strange She told me that my grandfather had died the day before and that my family was in Austin, Texas, for the funeral. I called Austin and my family was there so I got on another plane and flew to Austin. They picked me up at the airport.

Not much of a homecoming, but it was just one of those things. I mean, my grandfather had a long and full life, but I certainly wanted to see him again. He was always sending me letters and asking me about where I was and what I was doing. I tried to keep him informed because he was interested. He had served in World War I and was very interested in military history.

I returned home to Louisiana but I never felt comfortable there. I had gone to a small high school; there were only 19 in my graduating class. I haven't seen any of those people since. I decided to move to Austin and go to the University of Texas. Of course, I needed to work to pay for that, so I got a job at the Texas State Controller's Office, in the computer room. Computers were pretty much brand new back then. I really liked what I was doing so I ended up taking a lot of classes from IBM, and some of those were up in Poughkeepsie, New York. I sort of got my education that way. If I wanted to stay in that field, that was pretty much the only way to do it because they weren't teaching programming or anything like that at UT at that time. A few years later they were. By then I didn't need it. I never had any issues being a vet. Nobody asked and I didn't say anything. Nobody really seemed to care back then.

Home

I was living in Austin when South Vietnam fell and I remember watching it on TV. It was a process; it didn't all happen in one day. I do remember Xuan Loc was one of the places that held out the longest and I remember when it fell. I also remember watching the helicopters evacuating the American Embassy. I think pretty much all of us knew that, when we left, it was just a matter of time. The NVA was just superior to the South Vietnamese. There was an NVA regiment in the area where we operated. We had a lot of fire fights with them and I'm telling you, they were good soldiers.

I have some health issues now because of Agent Orange. I have some bad heart problems and I know that there was no history of that in my family. Many of my "brothers" have had serious health issues for years. For some, each day is a struggle. I am not talking about statistics, I am talking about men I personally know. It breaks my heart but their courage inspires me.

I had a friend that I met in Vietnam that actually grew up not too far from where I did in Louisiana, but we never knew each other then. He was in first platoon, and I was in third platoon, so we became really good buddies. And, as it turns out, he ended up working for the State of Louisiana, and he worked with my sister for 30 years. We never really did keep in contact after Vietnam. I went to Shreveport in January of 2007 to see my sister and my buddy, Donny Burg, was there. We talked, and talked, and talked. He had been to a reunion of our guys and told me there was going to be one that summer. He asked if I wanted to go. And I told him, "No." Finally, he talked me into it. We went and I enjoyed it. It was a little bit awkward, but I enjoyed it. Then on the way back after the reunion he said, "You need to go to the VA," and I said, "For what?" And then he said, "I think you got PTSD." I had been having some issues which no doctor had figured out, it was almost kind of like seizures. I told him about that and he said, "You're having panic attacks." I said, "Oh, bullshit, I'm not having panic attacks."

Anyway, I went to the VA and they diagnosed me with PTSD. They looked at my medical records and they said that, yes, I was having panic attacks. So they put me on some medicine and those seizures—attacks—just went away almost immediately. I haven't had one since. But, I have dreams. I can't remember them after I wake and I'm thankful for that.

In 2009, I had my heart attack. It was kind of out of the blue because I had just had a complete heart work-up six months before that. I was 61 years old and they said that a 61-year-old couldn't have a better heart than I've got. And these were private doctors, not the VA. Then I had the heart attack just out of the blue. Somebody said, "Well, you need to file a claim on that," and I did. So, now I have a service-connected disability for the heart issue. I am also service-connected for PTSD and tinnitus ringing in my ear. My ear is ringing 24 hours a day. And I have the scarring from the fragment wounds. Statistics say I will not live as long as non-Vietnam vets. I have lived on borrowed time for many years but I have lived life to its fullest. I am way ahead of the game and have no complaints. I have been blessed.

I have been to the Wall. I was almost emotionless when I was there. I was in Washington on business with a friend that I worked with and we decided to go there. We walked around the Mall and then, one day, we went to the Wall. It hadn't been there very long. I don't know. It was almost like I didn't know if I needed to be there or not. I guess it's the best way I can explain it. I haven't been back. And haven't really wanted to go back.

Looking back, to be honest with you it was in a way probably the best thing that ever happened to me. Because I grew up real quick, going from, you know, schoolboy to leading a platoon of infantry soldiers. And it was probably the worst thing that ever happened to me, just because of some of the things that happened. It's kind of like a Chinese sweet and sour dish. That's what I would call it, sweet and sour.

I met some great people in Vietnam that I'm in close contact with now. I went many, many years and didn't talk to anybody. Most of the guys I am in contact with are exactly the same way. They didn't reach out to find anybody after they got home. Now I'm in close contact with my old radio operator. He lives in Minneapolis. And I'm in close contact with Donald Burg, the guy I was talking about earlier. He still lives in Louisiana. I'm also in close contact with my old company commander. I talk to him probably once a week. And, you know, it's become a brotherhood again. I have a small extended family of "brothers." My love, loyalty, and respect for them and theirs for me are an important part of my life. We have peeked into each other's soul, and that kind of brotherhood is special and everlasting. I love them all and would die for them. Do not ever doubt that.

After many years, we got back together. I think we're all glad we did. But I don't think it would have worked out that well if we had tried to stay together right after coming home from Vietnam. I think everybody needed their own space.

I had some bad days in Vietnam and I've had some bad days since. I expect I will have some bad days in the future, but I have never had a day, an hour, or even a moment when my love for this country wavered. At no time have I felt this country owed me anything. No amount of military service can repay the debt I owe this country and nothing I have done or will ever do could possibly repay those who gave all. That's not just talk, it's a fact.

I still get chill bumps each time I hear "The Star-Spangled Banner." I never fail to remove my hat and if there is someone nearby that doesn't I will remind them. "The Star-Spangled Banner" means a lot to me.

It has been 45 years since I was in Vietnam and I still think about it daily. I suspect my memory is selective but it is the only one I have. A person's mind is complicated. The aftermath of combat is lifelong. It becomes part of who you are and you can't walk away or hide from it. You have to work hard to keep it from defining and controlling your life. I have learned fear, heroism, and death have a face and I have seen them all.

I now know every man has the ability to be good or evil, compassionate or ruthless, terrified or steadfast, and there comes a time in every war when each is appropriate. You do not choose those times, they just happen. You can question yourself until the cows come home but what is done is done.

We should not be teaching our children that everyone is a "winner" and everyone gets a trophy for participating. Visit a battlefield just once and you will only see "losers" and "survivors." There are no winners and no one cares about trophies. The world is real and there are plenty of bad guys who have no concept of fair play.

I am alive today because I was taught how to react without thinking. After all these years I still do that on occasion even though it is no longer necessary. This is sometimes a problem for me and those around me.

I was never the same after the war. Laughs were harder to come by and excitement harder to find. I never felt lonely but I felt alone, and there is a difference. Truthfully, I was fine with it and still am.

I never feared anyone or anything after the war and I say that out of bewilderment, not bravery or courage. It is not normal for a man to have no fear. You do not talk about it lest you be seen as a blowhard or braggart. A lack of fear is nothing to brag about; it is just weird and hard to wrap your brain around. We walk among you and I am not the only one.

I sometimes hear people say, "I feel sorry for 'John Doe', he saw too much war." Well, "John Doe" may feel sorry for you because you missed out on the war. And then again, maybe not. Only "John Doe" knows, and he ain't talking.

Specialist 5th Class George F. Pullen
25th Infantry Division

I joined the Army in April 1966, and served until April of 1969. In actuality, I joined to get out of the house.

I worked several MOS/job specialties while in the Army, including 71Q Military Journalist, 11C Mortars, and 71H Personnel Specialist. I volunteered for Vietnam duty and served with the 25th Admin. Co. Public Information Office, 25th Infantry Division, HHC, and Delta Co. 2nd Bn. 27th Infantry Reg. "Wolfhounds," and HHC 2nd Brigade Public Information Office 25th Infantry Division.

Coming Home

When I left Vietnam I thought about leaving behind the guys I knew. I knew the job wasn't complete, but I felt I contributed all I could and left my job in very capable hands.

I arrived at Travis Air Base and transferred to the Oakland Army Depot for processing. Then SFO for the flight home to Eugene, Oregon. All I can remember about the trip was that it seemed easy and uneventful. I think my mind was still occupied by the war and what was going on there. I was concerned for those still there, and I remember feeling we were fighting an impossible war to win. I mean it seemed like we were fighting with one hand tied behind our backs (free fire zones, no fire zones, etc.).

I was home but I was still in the Army. I had a 30-day leave before I had to report to my next duty station. I had called my family from San Francisco to advise them what time I should be arriving. I wasn't totally comfortable at home, but it was good to see my family and have some down time. My family was happy that I had made it home alive, and many said that I looked and seemed more mature than when I left. My family and close friends were good and didn't "pry" me for information.

As far as my friends go, for the most part, well, many were curious as to my year away, but I really had been away two years. Enlisted, basic training, AIT, and then being stationed at Fort Lee, Virginia, in the Public Information Office, so many of them had gone off to college and moved away from home.

I wasn't treated badly by the public. Only when I attended the Memorial Day Parade in Eugene, Oregon, in my uniform did I receive some negative comments, but I put it into perspective because Eugene, a university town, was very anti-Vietnam. I was actually interviewed by the local TV station about my time as a photojournalist, and it felt as if they were pushing me into making a political statement against the war. I was still on active duty so could not or would not comment on that.

One good factor was that I was returning to the army at Fort Ord, California, to finish my enlistment. The base was surrounded by communities that relied on the post for jobs and the money spent in those communities, and was insulated from the outside world.

Home

I experienced very little disconnect from my family, as I learned an appreciation for them and a respect for them I had not felt before. Remember, I joined the Army to get away from them. There was some disconnect from friends only because they had chosen a different path than I had. Eventually, I was able to reconnect with my friends. It took a while but at my 20-year high school reunion, we rekindled relationships. I had placed Vietnam memories, photos, and memorabilia in a couple of boxes and hidden them in a closet. I held no resentment toward those who chose not to go.

When Saigon fell in 1975, I was living in Portland, Oregon, working for a bank, married with a child, plus attending college full time in night school. I remember thinking, "What a fucking waste! All those lives, on both sides, what a waste. Why, why, why? What was the point? What did we accomplish other than killing a lot of people?" As you can see, I had more questions than answers.

I believe my "saving grace" was being able to put Vietnam in a box and put it in the closet for so many years. After 1975, not many people even spoke of Vietnam, period. No one asked, and no one talked about it, and because I was not from a "military family" they assumed I was okay. For the most part, I was.

But I had other issues to deal with that I had never dealt with until 1977. Vietnam was "nothing" to what was killing me inside. I was married now with two children, and I was gay.

I was in good physical condition and healthy most of my life, and it was a shock to find out I had high blood pressure, neuropathy, and skin cancer. I went to the VA clinic thinking they could give me answers and tell me what I might expect if these illnesses were Agent Orange related. I was advised by the VA clinic that these were probably not Agent Orange related, but that was two years ago, and I probably won't go back. I didn't go with the idea of getting anything but answers, but I was made to feel I should not be there. Honesty is vital for any business, and I feel my experience with the VA clinic in Vancouver, Washington, was not really on the up and up.

In fact, after my "last" appointment they made my next appointment for a year later. I had asked to see a dermatologist, and this was denied by my primary doctor. I thought, "I'm going to return to my private dermatologist to see if my suspected skin lesions are cancer." So, I was able to see my private doctor; within two weeks he took biopsies, and two of the four were cancer.

My wartime experiences made me a lot more aware of the plight of others to survive in this world and that I should reach out and help those less fortunate. It also made me improvise to do more with less, to barter for things your outfit needed to accomplish its mission, like the time we bartered two captured AK-47's for a photo dryer so our photos would be glossy and it made them easier to be picked up by the wire services, etc.

My experiences made me work with others under difficult situations, terms, and conditions, and to also be relied upon and rely on others—trust.

SP/5 John Swiderski
USARV

I was stationed at Long Binh and when I received my orders for home I went to a restaurant, someplace on post, with a few other guys who were going home at the same time. In order to clear post I had to take all my equipment and drop it off. They checked it to be sure it was all there, otherwise you're supposed to pay for it.

I'll tell you one quick thing: they had a 4' x 8' sheet of plywood, and everybody lined up. And you dumped all your stuff there and tried to put it in some kind of order. And there was a guy on the other side, with his check cards. And he's checking off the sheet, "And you got this and this. And that looks like it's all dented up." "What do you want me to do about it?" But I'm watching this—this guy is kind of nervous; this infantry guy. He looks like he just came out of the boonies: he has mud on his shoes and he is perspiring. When they were on base, they would usually take a green towel and put it around their neck, just to help them cool down, I guess. And this clerk is standing there. And he has his clipboard and he says, "Oh, you're missing this, you're missing this and you're missing this." The guy reaches across, grabs him by the collar and says, "You check everything off, right now, or I'm going to stab you." When he let go the guy checked everything off, silently, gave him his papers, and he walked off.

Specialist 5th Class Robert Hutcheson Collier III, (*aka* Hutch)

361st Signal Battalion, 1st Signal Brigade

I received my draft notice for the pre-induction physical while I was in college in 1969. I wanted to continue my education, if possible, so I joined the Army. My MOS was 32A10, Fixed Station Radio Attendant. I was trained at Fort Monmouth, New Jersey, and then sent to Vietnam.

I remember getting my orders. I was disappointed that I did not get a better TO&E assignment. I did not get billeted in my MOS and ultimately I never worked my MOS. I remember feeling scared and excited at the prospects of going to Vietnam.

As for my parents and their opinion of Vietnam, to this day I truly don't know how they felt. They did not openly discuss that with me, ever. My buddies were all over the map, some were talking about Canada, some were wounded returnees, very outspoken, pissed off at the government and angry at everyone and everything. I did not understand that until later. Then there were the guys like me, scared but duty-bound to do our country's bidding and protect people who believe in human equality and capitalism. I did not buy into the domino theory, I just knew that the South Vietnamese were in trouble and needed help.

I was assigned to Charlie Company, 361st Signal Battalion, based out of Qui Nhon. This would be my home for the next year.

Coming Home

We literally were the last four members of the 361st Signal, Charlie Company, out of the compound on the LST sand spit in Qui Nhon. We had two trucks, a deuce and a half and a 3/4 ton, and a bunch of equipment to turn in at Phu Kat, Tuy Hoa, and Nha Trang. We took the coast highway and sometimes side roads when there was activity going on around us. We had rain three of the four days on our trip. When we arrived in Tuy Hoa, we met and talked with General McCaffrey. Up to that point in all my experiences in-country, the "lifers" I had encountered were real pieces of work and not worth the powder to blow them to hell. This man was genuine, warm, caring, and actually redressed his full bird aide when the aide started in on us about the shape of our uniforms and that we did not come to attention when General McCaffrey came over to talk with us. Take into account we had been traveling for four days, sleeping in the trucks, eating at local stops along the way south. My truck had no top, not even a canvas top, we had rain, dust, and no showers, so we were pretty dirty and scruffy in the eyes of this stateside bird.

My year was over. It was time to go home. I was sent to Tan Son Nhut to catch a Freedom Bird home. I spent that flight with a high school buddy I went into the Army with on the "buddy basic" enlistment program. We played cribbage for the 24-odd hours it took to fly home, stopping in Atsugi, Japan, for fuel and finally landing at Travis Air Base in California.

We traveled from Travis to Oakland Army Depot by bus and received absolutely the worst believable reception. The left-wingers from Berkley were out in force with garbage and spit to welcome the "baby killers."

I was held at Oakland Army Depot for about 12 hours. My parents drove to Oakland from Boulder Creek to pick me up. I was jet-lagged and emotionally empty…I had no feelings whatsoever about being home except I was fearful that I was an animal and did not belong around people. I was in my home within two hours of release from Oakland Army Depot.

I was extremely uncomfortable in my parents' house. I did not know why but I was anxious and needed to get out. Most people in town treated me with curiosity but were indifferent to how I felt morally or mentally.

I got along okay, but my stepdad and my mother both commented many times in the days after I came home that something was very different about me, something that was good was not there anymore.

Home

I have never regained the close relationship I had with my mother. I was able to have a civil but not loving relationship with my father and my stepfather.

I started having nightmares. I could not remember what happened in them, only waking covered in sweat, shaking, and paranoid, listening for anything. Those problems and general anxiety got worse when Mediterranean fruit fly invaded California. Every night for several weeks they flew from three to seven choppers—Hueys—in formation, spraying bug spray over the Santa Cruz and Santa Clara counties. I became sleepless and extremely angry at everything. I just couldn't talk with my wife or my family, nobody could understand me, I couldn't understand what was happening . . . that is the time when I realized that there was really something wrong with me.

I remember when Saigon fell in April 1975. I was in my home, married to my first wife, watching the evacuation of the consulate in Saigon. It hit me like a truck. I began sobbing, shaking, and thinking, "Oh, my God, what have we done? Why have we wasted so many lives? What will happen to the loyal Vietnamese that we are leaving behind? We have broken our promise . . . fucking politics." I had worked closely with ARVN NCOs and officers. I lived with their families, and I was afraid for them. I feared they were dead or soon to be. I remained in an extremely pissed off mood that surfaced unexpectedly off and on for years to come. I was not able to get along with anyone in my family. I was on the verge of quitting my job. I hated my life and felt embarrassed that I did not give more to bring about success and give meaning to all of those who sacrificed it all. That was the first inkling that there was something wrong with me. I still did not get it. PTSD was a real thing, and I had it.

The war had a lot to do with the end of my first marriage. The most dominating factor was my inability to understand that she or any woman could even have my back to the death. That is way different from "until death do us part." There would never be the intimacy, not in a sexual way but a spiritual way, between her and me that men and women in combat have. This led to distrust, loss of the will to communicate and share what a couple must do to coexist in a relationship. Even children did not bridge that gap. In fact, I did and do have a better and closer relationship with my children than with their mother, but I can't explain that.

Not until 1985 did I seek professional help. Even then I was afraid to admit that I needed it. I was afraid that by admitting it I would lose my job, my friends, become a transient, or worse. I only got help long enough to deal with my worst triggers; smell and sound. I was working in a company producing electronic warfare and countermeasure components and systems, and I was afraid I would lose my security clearance if I "went to a shrink," so I opted for a civilian who happened to be a former Marine medic. This guy probably saved my life. He gave me the basic tools that I use to this day to cope with many of my triggers; the smell of diesel, two-bladed helicopter blade slap, sharp noises, night driving, crowds.

I finally reconnected with the VA in 2014 and now attend regular counseling. Sometimes it takes a long time coming home. If not for the persistent urging of fellow vets with the same issues I was trying to deal with, I still would not be home. I am finally finding peace in my life and a love for life that has been gone for a long time.

Vietnam changed me forever; PTSD, nightmares, anxiety, uncontrollable anger, paranoia in crowds, extreme emotional swings, health issues suspected to be related to Agent Orange exposure

In retrospect, I went to Vietnam because I believed in this country and what it stands for. I am saddened by the fact that so many died for no real reason, and that seems to be repeating itself over and over again. I am grateful to have known and worked alongside fellow G.I. and ARVN soldiers for a common goal . . . stay alive. I have also gained insight into how diverse people are and yet the same around the world. In my heart, I know we could all get along if religious and political beliefs were set aside. I still struggle with PTSD, and I have diabetes and heart disease attributed to my service in Vietnam. I can honestly say I have been forever changed by this experience: however, I would not trade the experiences that I had there for anything. Lastly, I would never wish those experiences on anyone.

I don't generally ever have a bad day. I think maybe that came from being in the jungle.

Sergeant Steve Kanaly

1st Battalion, 8th Cavalry, 1st Cavalry Division

I was drafted into the Army in 1966. There was a lot of us being drafted at that time. A lot of kids I knew in high school were also drafted. I was in college at the time but I dropped a class and, bingo! I was drafted. I know people that did everything possible to stay out of the draft short of going to jail. I don't know anybody that went to jail, but some managed to find medical means or other ways to avoid going.

I went to basic training at Fort Ord, California, near Monterey. I went in November and it was cold and wet. After you are inducted they give you a battery of tests and one of them involved symbols in Morse code. I have a musical background and that was easy for me. I'm sure I got 100% on it. And so they sent me to signal school, which happened to be also at Fort Ord. I did well, I was top in my class. So then they sent me to radio/teletype school and then crypto school. That school was at Fort Gordon, Georgia. Close to the end of the class they told me I would probably be going to Vietnam but they could stop that if I went to Officer Candidate School. They told me I would have another two- or three-year commitment and I told them that I would just as soon take my chances rather than do two more years in the Army.

The orders came down for my graduating class and in typical army style they were alphabetical. A thru K went to Vietnam and L to Z went everywhere else in the world. One initial, a life-changing event. What did I think about when I got my orders? Life sucks, I wish my name started with L.

My dad had served in the Navy during World War II and he hated it. He was stuck on an island in the Pacific in a repair unit. He worked on ships when they came into the harbor for repair in the Solomon Islands. And to hear him tell it was the worst thing in the world. But, to me it sounds like tropical paradise. He hated the heat but he was not getting shot at. I used to tell him that.

I think my father began a lifelong bout of depression when I went off to Vietnam. He never really recovered from that. It was always part of him after that even though I came back in one piece. I remember my mother told me later that the day I took off to go to report, he walked something like 25 miles in the middle of the night, and ended up at some friend's house, lost and disoriented.

It was in July of 1967 that I went to Vietnam. We flew to Manila and then changed planes. From Manila we flew to Pleiku. I stepped off the plane ready for some fresh air but it's not really fresh. The place smelled just like rot, jungle, burning shit, and I don't know what else. Everybody was burning shit and so the whole place smelled like that.

They put us new arrivals in some quarters for the night and later that night rockets came in while we were sleeping. I jumped out of bed in my skivvies and ran outside for a bunker and cut my foot on a piece of angle iron. But, some other guys never got out of the goddamn bed. They never got up, that was it. First day in-country.

That was my welcome to Vietnam. I said, "Fuck! I am going to get through this son of a bitch. Somehow, I'm going to do whatever it takes to get through this year and I'm going home." I don't know if it's about being brave, or being careful. You just do your job and don't go to bad places if you don't have to. Don't take unnecessary chances. I was pretty careful. I was assigned to the First Air Cav and I thought, "Oh, hell, great. Now I'm going to be in some unit that wants to go out and prove something." But it turned out that the Air Cav was a great unit to be assigned to. And, I did a lot less walking than a lot of other guys. I spent a lot of time in helicopters when other guys were just walking.

I was a radio operator in the 1st Battalion, 8th Cavalry. I was sent to a company in the field for about a month, but because of my Top Secret Security Clearance I ended up in Headquarters Company. We were a very mobile unit that traveled all the time. We were constantly on the move, always at forward bases and making new bases, clearing the mountains, and looking for bad guys. I lived in a hole that I dug. I would always be moving and digging another hole. Filling up sand bags and making a poncho roof.

My job in the tactical operation center was seven days a week, from midnight until noon. And, that was a very kind of odd thing. I never got to change it. It was always midnight to noon. When things are going on and action is happening all around you, sleep is impossible. I had to man the radio because my job was staying in touch with all the company radio operators. I had four that I maintained contact with throughout the night. I could actually take a power nap without lying down. I could sleep sitting up. I could almost sleep standing up. The squelch static from the radio would stop when someone would transmit. As soon as it stopped I'd be wide awake. And it would be something like, "We've got enemy moving around here. We got all kinds of shit going on. We need support right away. Can you hear me?" They would be whispering because the enemy was all around them.

I feel like I was lucky. All the real grunts were out in the field every night, my God. They had it tough. I feel pretty lucky. I lost 50 pounds in one year in Vietnam. We got stuck eating a lot of C-Rations which were pretty horrible.

We went in to reclaim an area called the A Shau Valley. This was a division-size move and was the biggest thing I ever saw. They prep this place for about three days with everything that could fly. I remember making this air assault in the first or second wave with 20 or 30 helicopters all together. We came over the top of this ridge into the valley and it looked like the friggin' moon. It was total devastation. Everything was knocked down. There were all these shell craters, some 20-30 feet wide, filling up with water; it was an amazing thing to see. We got on the ground, set up, and nothing happened. Nobody encountered anybody for about three or four days, and then they made contact with Charlie in the mountains. Rounds started coming in! From a howitzer that they think they moved to cross the border into Laos or Cambodia. I forget what it is. They started launching these rounds but they didn't have a forward observer so they were not very accurate. Every time a round would come in, they'd crank up all helicopters and just move them to the other end of the landing strip. Everything sort of fell harmlessly. What happened was, when they did all this bombing, the enemies moved out. They just left everything.

Coming Home

I was up at Quang Tri when I received my orders to go home. I went to Da Nang to catch my flight. I had a couple of war souvenirs that I was bringing home. I had a rifle and some other things that came from the A Shau Valley and, wouldn't you know it, somebody stole them there at Da Nang.

They put us on a commercial jet in Da Nang and flew us to Fort Lewis, Washington. There they gave us like a little debriefing, a good shower, something to eat, and a dress uniform. From there I took a flight from Washington to Los Angeles, where I was met by my family. They made me a great dinner, like Thanksgiving! I had a few drinks and ate about half the dinner before I pretty much fell over asleep in my plate. My mother says I slept for 23 hours.

My service time was not complete, as I still had time to serve before my discharge. I was a private pretty much most of the time I was in Vietnam because I was busted. Somebody found marijuana in my hooch and I wasn't even there. "But it was in your hooch," they said, "so you're getting an Article 15." So, I didn't get promoted until the very end. And then I got promoted two ranks, making me an E5 Sergeant before I left. So, now when I came back they said they needed a platoon sergeant. If I had still been a private, they would have let me out. So they sent me to Fort Hood, Texas. It was the most godawful, horrible duty. The only thing that saved me was I drove there. Austin, Texas, and the university was only an hour away so I enjoyed a real life away from the military. I went there almost every night. I made a lot of friends and it was the one thing that kept me together. I was very angry at being assigned to this place and this job. They should've let me out. I remember the day I was finally discharged, I pulled up in front of the gate, parked my car and yelled obscenities at the Army and said I'm never coming back here. Of course, I did television series in Texas for 13 years and did movies there, so and I have a lot of friends there now.

Home

When I came back I believe I was a different person after two years in the Army and a year in the jungle. I was more of a free spirit at that point. After I was discharged I got a job right away from my mentor who recommended me to a trap and skeet shooting club in Calabasas, California. It was something I could do three days a week and go to school. So, from the time I got back I was pretty much on my own.

Probably for the first ten years after coming home from Vietnam I would wake up suddenly and I'd be on my feet. I would just jump out of bed and scare my wife to death. I don't know if it was a specific dream or what it was about. I think it was just something held over from that first time they'd thrown rockets at me and all the other times after that. We were subjected to a lot of snipers and 122 rockets.

I remember I got in an argument with a soldier about some little bunker where I wanted to sleep. Finally, I said, "Fine, you take it, I'll find someplace else to sleep." So, I'm running the radio at two or three o'clock in the morning and we take incoming rounds and a rocket hit in that bunker where I might have been sleeping. One round and they died.

You know, I think waking up is about the rocket. I never sought treatment for this. I probably could have used some counseling. I just let it work its way out and eventually it stopped.

I enrolled at Valley State University Northridge as an art major. Between the job and the G.I. Bill I was able to go to school full time. School was hard at first. I didn't know how to start; I didn't know what to do. I took an art class, but it was an advanced class because I had a lot of other experience and classes but I had been away a long time. I had a wonderful teacher and he helped me get on track. By the end of the year I had finished 25 paintings and ended up having a one-man show. People liked my work. I sold paintings. I was a painter! That's where I found my way to express myself. I'd be outside the class with my easel flashing on these canvasses with big brushes and paint. It was a very physical kind of thing. I think there was a mystique about a Vietnam vet standing up there painting by himself. People either liked me or stayed away from me. I felt I was very welcomed in the college where I went. I'm totally different today, I don't paint like that at all anymore. Now I paint very controlled water colors, but then I was looking to find myself, I suppose.

I was 19 when I went to Vietnam. And, looking back at the whole experience, you're forced to grow up and be a man. You work hard, you work with others, and do so many things that you'd never done prior to that. And it made you a much more productive guy. I know I came back to college with a lot different attitude than when I went the first time. It made a big difference. Oh, jeez, you can actually get A's! Isn't that amazing if you only work at it? I had a lot of success when I came back.

While I was working at the gun club I met a lot of people in the film industry. Being a Vietnam veteran probably led to an opportunity for me to become an actor. I had never acted in anything before in my life. And, generally, I was kind of shy. My friend, John Milius, wrote the original screenplay for the movie, "The Life and Times of Judge Roy Bean," which was a western to be filmed in Tucson, Arizona. He got me a job on the film as an actor playing Whorehouse Lucky Jim. I had a one-week guarantee. I didn't even know what that meant. The cast was amazing—Paul Newman, Ava Gardner, and many more. It was directed by John Huston.

I drove John's car down there because he wanted me to bring our shotguns and his dog, so we could go hunting on our days off. It was about a nine- or ten-hour drive. And on my way down there I thought to myself, "What the hell am I going to do? How am I going to do this? What makes you think you can be an actor?" Finally, I said to myself, "Well, you know, if I screw up they're not going send me back to Vietnam. So it can't be too bad." I learned you sink or swim. So I just soaked it up, watched everybody, asked questions, and learned as much as I could, as fast as I could. I am an artist, a musician, and I love movies, so it kind of came together and was easy for me.

I ended up working ten or 11 weeks, the whole length of the picture. My part kept expanding because my friend, John Milius, was the writer and he expanded my part to the length of the film. My part is all the way through the movie and I made something like $1,000 a week, and that was huge money to me at that time, more money than I'd ever seen in my life.

I decided to take up acting and fortunately there was an actors workshop that was recommended to me, and that was the beginning of my acting training. Later, I had other training that was covered under the G.I. Bill.

I landed the roll of Ray Krebbs on the television series "Dallas," which ran from 1978 to 1991. Through my friend, John Milius, I was a technical adviser on the film "Apocalypse Now" in 1979 because of my experience with air assaults while I was with the 1st Cavalry.

I made one movie about Vietnam in 1989 called "Last Stand at Lang Mei." It was filmed in the Philippines. I played Major Verdun. I always wanted to make a movie about Vietnam.

When my television series was very popular, I was invited to attend the welcome home parade and installation of the Vietnam Veterans Memorial in Washington, DC. I was extremely moved to be a member of the parade and to be invited to the dedication ceremony for the Wall. A million people came out for this parade and ceremony. You could hardly get near the monument. I don't think I ever did on that occasion. There were so many people that came out to remember the fallen and to say "thank you." It was the real welcome home to me. Whenever I meet another Vietnam vet I always say, "Welcome home." And I always make sure to thank every veteran I meet for their service to our country.

It was a horrible thing when Saigon fell in 1975. Horrible seeing all these people just barely getting out and the place being taken over by the communists. I thought, "Jeez, what about the 58,000 Americans that died over there? What about all the Vietnamese that died? All the suffering and all the pain, and now it's just going to be a footnote in history." It seemed like an incredible waste. You knew anybody that was aiding the allied forces would be sent to re-education camps. They lost everything.

Vietnam was one of the biggest experiences of my life, no question about that. I was living without any conveniences. You learned that you don't need anything. You can get along with a small amount of water for two or three days. You're not going to die! You know, people tell me they're hungry. You're kidding me! Can you make it to the next exit? You don't know hunger. We had one mission where we were on top of a mountain. We were dropped there, and took these Bangalore torpedoes, and it blew down all the trees and we dug holes and we established a perimeter, and then the rain came. And the rain never stopped. It rained for like five weeks and only seldom could they bring in any kind of resupply. Everybody's rations were down to zero and we were drinking rain water out of our steel pots. I went three days without food there and I will never forget that.

Going through that experience was life-changing for me. It affected me the rest of my life. How I see other people. How I look at each day. How I appreciate things. Just being alive. I'm very optimistic because I know that I came back from there. I survived. I felt like I came back from having everything taken away from me to this opportunity where everything was available to me. I don't generally ever have a bad day. I think maybe that came from being in the jungle.

Captain Robert Madore
498th Air Ambulance Company
AVEL Central
520th Transportation Bn.

I received my draft papers ten days before my 19th birthday, which required that I report September 8, 1966. I was accepted into Michigan Institute of Technology, Sault Ste. Marie, and was scheduled to start September 15, 1966. All appeals for college deferment were rejected as I was unable to be in college before September 8. One week changed my life forever.

I was drafted with two of my friends from high school. We boarded a bus in Flint, Michigan, and headed toward Detroit where we were briefed and questioned if we had any medical abnormalities or were homosexual. All who passed, including me, were boarded on a train headed for Fort Knox, Kentucky, for basic training.

While in Fort Knox for the first week, I underwent thorough medical exams, inoculations, and a battery of aptitude tests. Apparently, I did well on the aptitude tests, because when they briefed me on the results, I was eligible for any school the Army had. I asked, "Which is the longest?" They responded with, "Field Radio Repair and Maintenance, MOS (Military Occupational Specialty) 31E20," which was 26 weeks long at Fott Gordon, Georgia. So, I re-enlisted for one more year to attend, which changed my status from US (draftee) to RA (Regular Army).

Well, basic started the next week. Winter hit, and it was very hard and cold, such that bivouac had to be canceled; even crawling through the obstacle course was canceled.

The two friends from high school had last names close to the beginning of the alphabet, so they were assigned to a different company than I, so I never saw them again until graduation day and then never since. I learned that they were assigned to Germany and other non-war zone areas for their enlistment.

Basic training was a snap. I was blessed with a drill sergeant who could not run, and a fellow recruit from Kentucky who kept the sergeant supplied with moonshine, so we did not have to run or exert ourselves very much.

While I was in basic, they herded us into a theater. There were about 500 of us there. This was for orientation for aviation in Vietnam. Well, I couldn't even spell helicopter! The sergeant that put on the presentation said that if anybody was interested in aviation, they should stay after the presentation. I was the only one that stayed. The next morning, they gave me a test on pattern recognition and a bunch of other stuff. That was all I knew about flight school at the time.

I graduated from basic and then went on to Fort Gordon, Georgia. The training for 31E20 MOS was so critical that they were teaching this course in three shifts. I went to school from 11:00 at night until 07:00 in the morning every day.

One day after school at 07:00 we had finished "dinner" and were on police call cleaning up the company area. The company clerk came over to me and told me I was supposed to be at battalion headquarters in 15 minutes for an interview for flight school.

I was pissed. No one told me about this. Here I was all dirty, unshaven, needing a haircut, and I was supposed to be at headquarters in 15 minutes in my dress greens with spit-shined shoes and all. I asked the first sergeant when he knew about it. He knew about it for two weeks. He never told me, because he did not like officers and figured he would get even with me.

I rushed to battalion and told them the situation and asked if I could reschedule. They said no. This was the only time they're going to be around. A major came out of the interview to get the next candidate, he was in his dress blues, and he tells me to wait around, he will explain my situation to the board, and they will call me last as this is the only time they will be here. Anyhow, my turn comes around. I am told to knock on the door, enter, and report. I report to the board. They're all in their dress blues. There's nobody there that's less than a major.

They asked me the first question: "What is an American?" and the questions got worse: there was no correct answer to any of the questions. I don't remember the answers to any of the questions. The interviewe lasted about half an hour. I found out later I was the only one picked to attend flight school. I didn't know that I was the only one until I graduated from my technical school. At graduation, they read out our orders, and my orders instructed me to go to Fort Wolters, Texas, for Primary Helicopter School. What a surprise.

I was able to go home for a short leave prior to flight school. When primary flight school was over, I was sent to Fort Rucker to learn to fly the Huey. The whole school lasted about a year from start to finish. I graduated in May of 1968, one of the top ten in my class.

At the graduation, they read off our orders. Now, we had heard rumors that the most dangerous job you could ever get was Dustoff medical evacuation in 'Nam. Well, when they came to my name and they read my orders to be a Dustoff pilot . . . the best way to describe how I felt was . . . I felt like I had jumped 20 feet behind and had left my boots right where they were in formation. That was the feeling. I went, "Holy hell, I've got the most dangerous job there is."

I was assigned to the 498th Air Ambulance Company out of An Son, which is about 12 miles west of Qui Nhon. They told us that after about 280 hours they would consider us for aircraft commander. I became an aircraft commander after about 28 hours. They gave me a bunch of maps, and I started my first tour as an aircraft commander.

I was returning from one of my night missions and at midnight I turned 21. Business continued as usual until I was shot down and wounded in May 1969. I was on a mission to support the Korean Tiger Division. Well, it was in a boxed canyon, I was flying, and just before landing, all hell breaks loose. AK-47 armor-piercing rounds hit the aircraft. It was later discovered that 20 entered the cockpit and 15 the inner compartment, and one had hit my legs, and one took out the radios and the intercom so I couldn't tell my pilot I was shot. They all came from the pilot's side, somehow missing him. They missed the whole crew except me. Without the intercom, I couldn't tell anyone I was hit, so I had to keep flying.

I landed at an ARVN artillery outpost about three miles away. I told my medic I was shot, and he wanted to carry me out. I told him, "No way. Get away from me. You can help me out of the aircraft, but you're not carrying me." Then the medic looked at my wounds and said my legs looked like a hamburger. Everything in the aircraft was leaking that could leak.

I was in the hospital for six days, and I had enough of the hospital. I told the doctor I wanted to go back to my unit. They wanted to evacuate me to Japan, and I said "no way." Well, they were going to anyhow, but I convinced them to let me stay. The doctor said, "Well, you can't walk." I said, "Yes, I can." He said, "No, you can't." He made me walk down the hallway, about 300 feet down the hallway, 300 feet back. He said, "If I see any signs of pain, I'm not letting you out of the hospital." Boy, that hurt like a son of a bitch but I did it.

The Doctor said, "I can tell you're in extreme pain, but you're doing everything you can to hide it. If you want to go to your unit that bad, I'll let you go." So I went back to my unit and recovered there for five weeks.

The flight surgeon finally cleared me for flight duty. The 498th had a policy that before I could assume the duties of Aircraft Commander again, I had to fly with another aircraft commander as pilot to demonstrate I could perform the duties. The first mission I flew was a hot mission; it felt like my skin got four inches thick and just came off, I was so scared. They were shooting at us as I was coming in, and all I saw was the wrong end of the muzzle, but I kept on going anyhow. That's how crazy I was.

From that point on, I went back to flying full time. I was an aircraft commander again, and I continued flying. I flew about ten months out of twelve, and I evacuated over 1,500 men.

Coming Home

Getting shot did not affect me. It was just another day, no different than so many others. My orders came through, and I went back to Bien Hoa where I started a year ago for the long trip home. I flew to Okinawa, and then Seattle, then went to the east coast to Lakehurst, New Jersey.

We were bused to Philadelphia to catch a flight home to Los Angeles. They warned us to get out of our uniforms as quick as we could because of the protests. After I had arrived at the Philadelphia airport, I changed into civilian clothes but not before some people spit on me and called me names. I tried to ignore it. In fact, I ignored most of the comments because it would have ended up in a fistfight. And I did not need that. I just wanted to get out of there as soon as I could.

My parents picked me up at the airport in Los Angeles and welcomed me home, but at that time I felt like I was just visiting. It was just a numb thing for me. I did not really feel like I was home.

When I came home, there were people shooting at cars on the freeway. It was a very dangerous time.

Because my parents had moved, I had no friends living nearby. I knew that the antiwar movement was huge in California at that time. Yes, it was pretty bad. I just tried to ignore it, and I guess I did a pretty good job because I really don't remember a whole lot. Looking back now, it is a blank spot in my memory. Things took place that I must have blocked out.

I was sent back to Vietnam in July 1971, as Executive Officer of AVEL Central. After two-and-a-half months I was reassigned as Assistant S-3 at Headquarters, 520th Transportation Battalion. I didn't do a lot of flying my second tour. I came home in July 1972 after my second tour. The second time I came home was uneventful.

Home

I spent seven years, three months and fourteen days in the Army. Now I was out. One of the things I remember about this war: the first time I came home was the day they landed on the moon. I was able to watch Neil Armstrong step out and put his footprint in the sand the night I got home from Vietnam. There is a lot of history that I just did not pay attention to.

I don't hold any animosity towards anyone. There are some things that bother me. Like the time I evacuated a young soldier. He had been a victim of a land mine and had both legs blown off at the knees. He did not know the extent of his injuries, as it is customary to cover him with a blanket and place two boots at the end. My medic told us over the intercom that the soldier had told him that he had the "million dollar" wound and was going to go home and that a football scholarship was waiting for him.

I don't remember a whole lot about my second tour. I remember my time in Vietnam primarily from my first tour.

Right now I have Agent Orange issues. I have diabetes. I also have glaucoma, and I lost my vision in my right eye. They are now attributing the stroke I had in 2013 to Agent Orange. The only thing I can be thankful of is I do not have the cancers that a lot of people have.

I would do it again if I had to. I would probably do it differently. I would spend longer time on the first tour than I did, but I was happy to get out of there with my life. I went home from that first tour with a friend from my company. His name was Roger Opendall, and we were both on the same plane. We both had the same sensation: Did we do everything we could? Why were we alive?

I still ask this question today. Why am I still alive? It was like it was mind over matter. For those people that believe in God, who are very religious, they always ask me did I pray before a mission. No, I did not. I did not pray that God would intervene because God gave me all the powers that I needed already. As for divine intervention, everybody who prayed for that is dead. They had given up on themselves and asked God to save them.

Specialist 5th Class Kirk Cooke
125th Air Traffic Control Bn., USARV

I was discharged there at Oakland. Here I was, an E-5, 13 ½ months in Vietnam, only 20 years old. I can't even buy a beer in the United States. So I grabbed my gear and caught a cab to the airport where I caught a plane to Detroit. I guess I was just feeling kind of numb at this point. I landed at Detroit Metro, wearing my dress greens, and was met with rude stares. No one said anything to me or spit at me; they just glared at me. I had an immense feeling of being let down. I thought, "Why do you hate me? What did I do to you?" I didn't expect that. I grew up with John Wayne and World War II. We were the good guys, you know?

I came back to an unappreciative nation. I was treated pretty much like dirt. I wore my uniform home, and people said, "Don't put it on again, they'll spit on you." That was the last time I wore my uniform. You are supposed to be proud coming home. You served your country; you did your duty. I did what the country asked of me, and, I mean, it still affects me.

I did not settle down for at least ten years. I got divorced pretty quick and then it was, you know, drinking. Then it seemed like I wandered, and then I got married again, then another divorce. I was just too angry, and there was no help. There also wasn't any thrill; there wasn't anything dangerous anymore. No adrenalin.

Specialist 5th Class Marvin York
AVEL Far North, USARV

I was drafted into the Army when I was 20 years old. When I took my physical I failed the hearing test. I failed it twice. They sent me to a specialist in Chicago (I was from Lockport, Illinois.) They even hooked me up to a lie detector. This time I took the test, I passed. They needed a body. They took me. Funny thing is when I was discharged the graph was the same as it was when I was drafted.

I turned 21 in basic training at Fort Knox. After basic I was sent to

Fort Monmouth for training as a backpack radioman with the infantry. That lasted one day. I was sitting in class, and a sergeant walked in. He said, "I need two volunteers for 26D20." I had no idea what 26D20 was so I happened to have a manual and flipped through the pages, and I saw that 26D20 was the MOS (Military Occupational Specialty) for Ground Control Approach Radar Repairer. Six months school! I volunteered for that, and he picked me.

I finished the class and graduated. I was promoted and saw there was another class available. I thought that maybe if I go to this other class I'll get another promotion, so I volunteered. So, the sergeant major of the post called me in and said, "York, you're a smartass." And I said, "What do you mean?" And he said, "Well if you got this class you wouldn't have enough time left in the service to go to Vietnam." He picked up a packet on his desk and said, "Here're your orders." That was it. I had two weeks leave and off I went to Vietnam.

I didn't have any thoughts about the war. I didn't care about it. It didn't bother me. It didn't matter. I did know people that were drafted, high school buddies of mine. But I'm a crazy kind of guy. I just didn't think about it. It didn't bother me. So when I was drafted, I didn't shy away from it.

I flew to California with a guy named O'Rourke who I graduated from AIT with. We were flying military standby and the airline put us in first class. We drank like thieves and ate ox tongue and cheese. Finally, they cut us off because we got too drunk and that's when the flask came out. I said, "Hey, give me some ice cubes in the glass," and we did that. They had to unbuckle us when we landed in Oakland. We couldn't figure out how to unbuckle the seat belt! We took a cab to Oakland Army Base and reported in. The sergeant looked at us and said, "Oh, Jesus." "We're toast," I thought. I said, "Here're our orders." "Get in the bunk over there. Go to sleep, you're flying out tomorrow." We had so much fun; it was just ridiculous.

I was assigned to AVEL Far North, an avionics company at Red Beach near Da Nang. That is where I spent the next 12 months.

Coming Home

Our major was giving early outs, maybe seven to ten days instead of having to spend the full year. My brother was going to Guadalajara with his family and I wanted to meet him. I told the major the dates and all of a sudden he changed his mind and canceled early outs, so I missed meeting them down there. But I flew home on my scheduled DEROS date.

I flew out of Cam Ranh Bay to Fort Lewis, Washington. The plane had a problem, and I don't know what the problem was, but we stopped in Tokyo. They wouldn't let us off the plane, probably because they were afraid we'd leave, and I would have! I was out of the service as far as I was concerned. I wanted to take a boat across the ocean. I wanted to get out, but they wouldn't let us off the plane. We finally left Japan and flew to Washington.

The guy sitting next to me was in school with me in Fort Monmouth. He was nervous and everything. I said, "What's the matter?" "My wife is going to meet me." "That's good." I wasn't married, I was single. "What's the problem?" "Well I've got gonorrhea" and I said "You *what*?"

I looked at him and said, "You're married?" "Yeah, my wife's going to want to . . ." "I know, I know, your wife is going to want to have sex, what are you going to tell her?" He said, "I don't know what I'm going to do." He was desperate. I felt sorry for him. How could you do that? You knew you were going home in a month, you could have gotten it treated. He obviously had done something real close to the time of leaving. I don't know what to tell you. What's he going to tell his wife? That's terrible.

I had my exit physical and left. I went to Lompoc, California, because my buddy, Van Schuknecht, who I called Shitty, had extended his tour for an early out and wouldn't be home for another six months. Yeah, we were best buds for the whole year. I went to visit his mom and dad and stayed with them awhile.

While I was there Van's sister, Kathy, was in college at the time and took me to a party with some college kids. So I went and wore my dress green uniform. I talked to everyone and I was having a decent time, and this red-haired girl, a real pretty girl, looked at me and says, "You're one of those baby killers," and I looked at her and said, "What?" I looked at her and said, "I'm not a baby killer, I've never killed a baby in my life." I don't know why she did that. I turned around and looked at Kathy and said, "I've got to get out of here." She says "Why?" and I said, "Because someone's gonna get hurt. Might be me, might not be me." So we left.

I spent a couple more days with his parents. Then I went down to see my Uncle Henry, George Henry, my mother's brother. He was a very talented man. Nice person, anybody would love him, anybody would. He could really play the piano and never took a lesson and couldn't read music. He'd play bar tunes and stuff. He was so much fun. That's where I took my dress uniform off and put them in the burn barrel and burned everything I had with the military. I just burned everything. Then I went home.

I was really looking forward to seeing the people I lived with my senior year of high school. I looked forward to seeing my twin brother, my brothers, looking forward to seeing my mom. I wouldn't call it apprehensive; I just didn't know what to expect. They hadn't seen me for a year; I hadn't talked to anybody for a year other than letters. I really can't remember the day I got home. But, everything was good. Then I found out my brothers had sold my '55 Harley that I totally rebuilt in 1965 or 1966 before I was drafted, and it was gone. My Harley was gone. I guess they needed money to party and drink. They sold my guns, my pistol, and my rifle. I didn't have any clothes. It was all gone. I didn't know what to do, so I left. I rented a small trailer and stayed there three or four months by myself, just thinking about things. Whether I wanted to get back into society, I just didn't know. It was too much. I finally got over that.

Home

I went to work for my brother, Dick. He had started a machine shop in his garage in 1963. He went from a garage to a 4,000-sqare foot building. One day my brother Bill stopped by. I was operating the cold saw we used for sawing material, a good size saw. I'm sawing material and my brother Bill set off a cherry bomb close to where I was, and my reflexes hit me and I just dove under the saw and hit the concrete. I turned around, and they were all laughing, laughing at me. I looked at them and thought, "You sons of bitches, you don't know what it was like in 'Nam. When a mortar hit, you dove!" That was my instinct; I just hit the concrete. I was very bitter at the time. They didn't do it again. I wanted to hurt them, but I didn't.

I used the G.I. Bill and went to junior college and took electronics. I started having nightmares. I drank heavily, really heavily.

That's where I met my wife, Gale. After I met Gale things changed a little bit. I had fun and really enjoyed our courtship. Her parents liked me, and I liked them. It was good. For some reason as I got older, in my early 60s, I started having nightmares and she was afraid to touch me or wake me up because I was moving around, legs were moving, she didn't want to touch me because she knew I was a combat veteran and she was afraid I might hurt myself or her. I told her I'd never hurt her and she could wake me up. She said she was scared to.

I'd have nightmares you couldn't believe. I thought I was under attack and couldn't get away; I thought someone was going to kill me. She asked me what I was dreaming about. I'd tell her I couldn't remember. I never told her anything about my Vietnam experiences. Never told my children. Never told anyone about it.

I finally went to the VA for help. I was in my 60s. I can't tell you my exact age. I was turned down at the VA in Atlanta, Georgia. They classified me as 8F. My buddy went to Asheville, and he got accepted. He wasn't in 'Nam, he was stateside all the time. He was an officer, an optometrist. Easy duty for him. He told me to go to Asheville, so I sent an application in. I forgot to include my DD214 form so they sent me back a letter saying, "Please send your 214 form in and answer, 'Were you ever in combat?'" Well, I wasn't infantry but I consider mortars and being shot at as combat. Whether I could return the fire round doesn't matter, I considered that combat. I just wrote a letter back with the dates from the beginning of my tour to the end. I was in combat. They didn't write me a letter back, they called me and said, "Can you come next week?" They classified me as Group 5. I went to the VA and got taken care of. I thought they were wonderful; they treated me right.

I hated the war. I hated the government, and I hated the American people for killing people, all the crap while we were in combat. It was not good. I just hated what they did. As far as my experience in 'Nam, I wouldn't trade it for the world, for the people I met, the friends I met, and the experience of being with men in war. That's how I feel about it. I wouldn't change it for the world. I loved the experience; it was dangerous, it was a high. I just wouldn't change it.

Specialist 5th Class Larry Butcher
HHC Company, 2nd Battalion, 14th Infantry, 25th Infantry Division

My father and grandfather were 100% anti-military. My father had been drafted but for some reason during basic training he was given a less than honorable discharge. The reason was never discussed.

When I was 17 years old my grandfather bought me a brand new 250CC Suzuki X-6 Hustler motorcycle and told me to leave the country, and I did. Two days later I was in Mexico sitting alone, thinking, "This isn't how I want to live my life." I returned home and took the motorcycle to my grandfather and told him he could have it back as I hadn't kept my part of the bargain. He told me to keep it and enjoy it because I would "die" in the Army. I was drafted and went in as a CO (conscientious objector).

I was inducted on May 1, 1969, and released on November 10, 1970. I was drafted just before the lottery took effect. My total active service was one year, six months and ten days. My MOS was 91B20, Combat Medic. I was married at the time I received my orders for Vietnam. For the first two days after I received orders, I didn't talk to anyone and couldn't eat. In my mind, it was a death sentence.

A classmate of mine from the fifth grade through high school was drafted with me. We went thru basic, AIT, and a 30-day leave together. We both reported to Oakland on a Friday, but we went AWOL. We checked in at Oakland and then his mother picked us up and took us to her home in Modesto for the weekend. Late Sunday we returned to Oakland and went straight to the chaplain's office. We told him we just wanted to spend the weekend at home. He gave us his business card and told us what building to report to and told us to say *only*, "Call the Colonel." We reported in, and they were really angry with us but could do nothing. They finally issued us our gear, and that was the end of it. We left the next morning for Travis AFB for our flight to Vietnam.

In Vietnam, I served in III Corps, headquartered in Cu Chi with HHC Company, 2nd Battalion, 14th Infantry, 25th Infantry Division. I was also attached to Delta Company for six months in 1969-70. Our area of operation included Dau Tieng, Tay Ninh, Go Dau Ha, Trang Bang, Ho Bo Woods, Boi Loi Woods, Cu Chi, Trung Lap, FSB Devin, FSB Pershing, FSB Patton II, Patrol Base Hunsley, Katum Airfield, and The Mushroom.

I went to Hawaii in June 1970 to meet my wife on R&R. When I got off the bus and walked through the lines of people, no one met me. I thought maybe she didn't recognize me after spending seven months with the infantry. I walked behind each line checking each face one by one . . . she wasn't there. I called her father in California, and he said there was a problem, and I should ask the Red Cross to call him. I waited 24 hours for the Red Cross to confirm what the problem was so I could get a compassionate leave. After the 24 hours, I went to the officer at Fort DeRussy and asked to be on the next flight back to Vietnam as I was going crazy waiting. The officer told me to sit down, he would make a phone call. He called my unit in Cu Chi, and they approved a three-week compassionate leave for me. The problem was my wife had taken the time off her work and had gone on a vacation with her new boyfriend.

While on that leave at my mother's house, a neighbor came over and asked me why we were over there killing innocent women and children. I just told her she wouldn't understand and didn't say anything else about it. Half of my platoon had been blown away by two mama-sans just outside of Trung Lap while doing a morning road sweep. How could I explain that to her?

While I was on compassionate leave, I lost my travel orders. To save money I registered at the local junior college and flew to Hawaii as a student, half price. In Hawaii, I hitched a ride on the first plane out that was going to Vietnam. It was going to Da Nang. In Da Nang, I waited overnight and caught a ride to Tan Son Nhut where I located my fatigues. Back in uniform, I caught a convoy ride up to Cu Chi. When I walked in my unit, everyone was shocked I had returned.

July 10, 1970, was my 21st birthday. I was in Cu Chi Base Camp in the NCO Club, celebrating, when we got the news. A Chinook had been shot down, and grunts from Delta Co. were on it. Two of the aircrew were killed and five of our grunts. Thirteen more were injured. Three Delta Company grunts I had humped with died as result of that crash. A fourth survived and was taken to the 12th Evac Hospital in Cu Chi. I went and visited him in the hospital and he told me his experience in the crash and how he was able to escape the burning Chinook. When our visit ended, I went outside and was physically sick.

I extended my time in Vietnam and was there one year, one month, and ten days. At that time the 25th Division was brought back to Hawaii, so I got an early separation from active duty.

Coming Home

I was scheduled to go home but, due to the 25th Division returning to Hawaii, there were many guys ahead of me. I went to roll call three times a day for six days before my name was called. After we had cleared the installation at Cu Chi, we were loaded on three buses. As the buses started leaving, a G.I. behind us shouted, waved his hands, and stopped us. We were told our plane had engine trouble and was under repair. We were taken off the buses and stood or sat on the dirt behind a chain link fence for 24 hours. We had guards around us, and they brought us C-Rations and water, but we had no blankets, cots, or chairs. As we boarded the plane and during the entire flight I hoped the mechanics hadn't overlooked anything when making the repairs. We landed in Guam and Hawaii to refuel then touched down in the morning at Oakland International Airport. When we landed in California cheers went up, and there wasn't a dry eye. Even the stewardess had tears. I will never forget that moment. We landed at Oakland International Airport as Travis AFB was fogged in. Oakland wasn't prepared for us, and they had us wait on the plane several hours before customs officials could check us in. We were then taken the short distance to Oakland Army Base for clearance. That same day, at midnight, I walked out, relieved from active duty because of my 180-day early release.

I lived in Modesto, California, at the time which is about a two-hour drive east of Oakland. Friends brought my mother over to pick me up that night. On the drive home, I was sitting on the passenger side in the back seat and could see the speedometer. The friend was driving South on I-5 at 100mph. I was thinking, "Damn, I lived through my tour in Vietnam, and I'm going to die a few miles from home!"

When I returned home, friends were talking about events I didn't know anything about. I didn't even know about Apollo 12. Many times I sat quietly while my friends and family were talking about current events because I couldn't contribute anything to the conversation. They weren't interested in Vietnam, and I didn't want to talk about Vietnam. I felt distant, numb, and very alone. During my tour "The World" and I had changed, and I was quite upset I had "lost" a year of my life.

Before I left Vietnam, I had my mother prepare a place for me to live when I got home. I didn't want to stay at my mother's house. Many nights I would wake up and be afraid to move or open my eyes. I wasn't familiar with the smells and sounds and didn't know where I was. I left a light on in the room while I was sleeping for several months.

I moved to Merced, California, in 1976 and bought a new house on a cul-de-sac. Castle Air Force Base just north of Merced was a B-52 training base before and during Vietnam. There were three B-52 pilots and a forward air controller pilot that flew light aircraft slowly to maintain constant aerial surveillance in Vietnam living next to me. They had many block parties and, as a Vietnam veteran, I was always invited. Having been a combat medic with the infantry out on ambush patrols and search and destroy operations, walking on the ground seven months, it was interesting to listen to their stories and how at risk they felt when they were flying their high altitude bombing missions. In the beginning, I had a difficult time relating, and the forward air controller pilot and I talked about it. He flew his missions at low altitudes and had difficulty relating too. After living there many years, I came away understanding we were ALL at risk. We all needed each other to do our part to come home alive.

No one wanted to know anything about my experience in Vietnam. So, I clammed up and buried it. I just tried to go back to "normal" and deal with it on my own.

Home

As a child I attended a Seventh Day Adventist private grade school and high school. I went to my 20th reunion and 40th reunion. I felt no connection with them both times. I felt like a stranger in their presence. I felt we had nothing in common and I won't attend future reunions. I don't know if that has anything to do with my service in the Army and Vietnam. I have never seen in all the years since that classmate with whom I was drafted, did basic, AIT, and then flew to Vietnam with. I do know he was wounded.

There are other horrible things I witnessed in the Ho Bo Woods I don't discuss or talk about with anyone. When I do think about them, I wonder why and I ask myself if there is something "wrong" with me. Why do I think about them and remember them so clearly?

Last year I was tested and received hearing aids from the VA. My hearing loss was deemed service related, and I received a partial disability. I went to an appointment with a VA psychologist August 18, 2015, and was pleased with the process. The psychologist has done extensive study and experience with PTSD. She has been deployed several times with different units and branches of the military. She was part of their chain of command, stayed with them, and returned with them. Today's military returning experience is very different than the common Vietnam veteran's experience!

Starting out, she carefully explained the process, her questions, and the reasons they were important, and then asked me what I wanted to tell her.

I told her about things that happened in Vietnam while out with the infantry I haven't told anyone . . . things I would never talk about outside a professional environment, things that after 45 years still haunt me and simply won't go away.

As I was standing, ready to leave, she made the comment; "You definitely have PTSD. There is no question about it."

I walked out feeling I had just taken one of the hardest examinations I have ever sat for . . . but I was only telling it like it was . . . there were no wrong answers, nothing to lose, no penalties, no peer pressure, no grade to strive for, no reason to be stressed out.

I sat in the car for a while listening to music before driving home. I remembered a comment my cousin's husband made to me; "Be Army strong"! And I thought to myself; I AM Army strong!

Agent Orange health issue? Yes. On February 7, 2015, I drove myself to the Mad River Community Hospital in Arcata, California. I walked in the emergency room and said, "I think I need to be checked out." I was immediately taken by ambulance to the main hospital and underwent an emergency double bypass. Heart surgery. I had a left anterior descending artery blockage, the one they refer to as the "widow maker." This blockage has a survival rate of between 5% to 10%. In the ICU my heart surgeon asked why I waited so long before going to the hospital. My answer was, "I was a combat medic attached to the infantry in Vietnam going out on ambush patrols and air assaults. I'm tough. After Vietnam I can take it, I can handle anything."

Like most, my wartime experiences changed me. I was concerned when I was on the Freedom Bird returning home at the end of my tour: concerned I had lost my ability to "feel" emotions. A WWII veteran that had been a POW in Germany told me, "It will always seem like yesterday, it will never go away." At the time, I thought the memories would fade and be forgotten, but he was right. Today, 45 years later, the hair on my arms stands up when talking about some events.

Everyone that served in Vietnam changed in some way. We each had different experiences, and we each dealt with them in different ways.

For me, personally, it was not possible to stay the innocent, carefree person I was after having watched other human beings shot and killed. It was not possible to stay the carefree person I was having treated my combat-wounded friends and buddies. There are 13 names on the Vietnam Veterans Memorial Wall I personally went out on ambush patrols, search and destroy missions, and air assaults with. I think about this with a heavy heart and a bit of guilt. I have always felt I should have been better trained, better prepared, and done more for the grunts I was responsible for.

For 45 years I have considered my four divorces, occasional panic attacks, and my waking four and five times every night as "normal," and not at all associated to my time as a combat medic out with the infantry in Vietnam. When I walked out of the Oakland Army Base at midnight, November 11, 1970, I was consciously very aware I had changed and was a very different person than I was before my tour in Vietnam. As of today, I have attended meetings with my VA counselor for three months and now realize many things I have always thought were "normal" are not. In Vietnam, my duty was, first, to keep my platoon healthy and, second, alive when they were wounded. To this day I feel I came up short treating my buddies that lost their lives in the boonies. I still re-live, second-guess, and question myself, wondering if I could have—should have—done more to save them. Other horrific events I witnessed out in the jungle have never been talked about with anyone until now, in this "professional" setting. Intrusive memories of events that I wish I could "un-see" and "un-remember." I have always wondered if there is something "wrong" with me for not being able to stop these memories from returning. I'm learning "Coming Home From War" never ends. Each of us has brought home experiences and memories that have forever changed us.

Looking back, no one could be prouder of serving in the Army in Vietnam than I am. I received two Air Medals, an Army Commendation Medal with "V" device and Oak Leaf Cluster, a Bronze Star with "V" device and an Oak Leaf Cluster. The medal I'm most proud of is my Combat Medical Badge. Having been out on ambush patrols and air assaults with the Army Infantry fills me with pride. We were outstanding soldiers, soldiers our country could be very proud of. We were sharp and did our job as professionals. I'm filled with sadness and a little anger when I think about how our leaders let us down and how disrespectfully our citizens treated us when we returned.

This morning I looked at my reflection in a photograph taken of me at the Wall. I realized a part of me is still with them. While I was looking at the photo my wife asked, "What's wrong?" I pointed to my reflection on the Wall and had tears. There are three on the Wall I cared for in the woods. One that died instantly and two that were alive when the urgent Dustoff left but didn't survive their injuries. I'm still haunted by the memories and wondering if I could have done something different or done more to save them. I am, myself, along with many others, still in the process of "coming home."

Specialist 5th Class Pete Poirier
AVEL Far North
USNS Corpus Christi Bay

Duncan, one of my younger brothers, was killed at age 19 while serving with the Coast Guard. Larry, a second younger brother, died at 27 after serving in the Navy. Neither brother ever had to go to Vietnam. I still, to this day, can't adjust to the fact that their lives were cut so short. I wished so much more for them.

I enlisted in the Army after getting a draft notice because I wanted

skills training, something that would be useful when I got out of the Army. I turned down OCS and helicopter school because I thought, at 19, those were slots that ought to be filled by people who were older and wiser than I was at the time. So, after basic training at Fort Benning, I was sent to Fort Gordon, Georgia, where I trained as an Avionics Navigation and Flight Control Repairer, MOS 35M20.

I was not surprised when I got my orders to go to Vietnam in 1967; I expected it. I originally was assigned to AVEL North, 4th Platoon in Chu Lai in 1967. I was in Chu Lai for the 1968 Tet Offensive. After that, I went to I Corps and helped establish AVEL Far North, 1st Platoon at Camp Viking, which was at Red Beach north of Da Nang. We slept in tents inside a sandbag and wire perimeter and had alternating guard duty where I manned an M-60. We routinely received mortar fire and hits from 122mm rockets followed, from time to time, by incursions on our wire. Most of our time, however, was spent in the relative comfort of an avionics maintenance and repair shop which we had set up on the PSP flight line. My second tour was seaborne on the USNS Corpus Christi Bay with A Company, 1st Transportation Battalion in 1969 and '70. Corpus Christi Bay originally was the USS Albemarle, a seaplane tender later loaned to the Army and refitted to become the only Floating Aircraft Maintenance Facility (FAMF) in existence. The ship's civilian crew all were in the Military Sea Transportation Service (MSTS). I was in the avionics shop doing major repairs on navigation and communications gear.

Coming Home

My final trip home started early because one screw on the Corpus Christi got bent when it became fouled with a cable. We sailed from Cam Ranh Bay across the Marianas Trench, which is the deepest part of the ocean, past the Philippines to Sasebo, Japan. Dolphins and whales swam in our wake while the ship lurched with each turn of the screw. We even had a huge (I mean gigantic) sea snake wash on board one night during a severe storm. The last leg home was a long flight from Japan to Seattle-Tacoma. The night before I left, some of the guys took me out for a separation celebration that made the flight a lot longer than it needed to be. I slept most of the way home.

When I got out of the Army in 1970 after two tours in Vietnam, the world seemed very different than it had been only three years earlier. My short hair and clean shave told everyone around me that I was different. My homecoming felt uncomfortable except for my immediate family. Old friends and acquaintances treated me like I had some communicable disease. They asked things like, "Did you kill any babies while you were there?" I would have preferred, "Man, it's good to see you safe and sound, so glad you're home."

I was devastated when I saw in real time on TV national guardsmen march up a hill, turn, and fire live rounds at students at Kent State, killing four. "Those aren't soldiers!" I shouted. That is probably the moment when I put my military service completely behind me and left it there for the best part of 40 years.

Home

I got a pretty good job, working for a computer utility, got into a pretty good school (Case Western Reserve), and hooked up with some good friends (all Vietnam vets). We rented a 28-room mansion on Lake Erie and became famous as party central for most of Cleveland. The truth is I burned out quickly, trying to go to school and work full time while becoming famous or perhaps infamous as a party dog making up for all that lost time in Vietnam.

My buddies and I decided to chuck it all and move to Hawaii. That worked well for a while, but soon the guys started to get "rock" fever and the group drifted apart. I stayed on, but after three years in Hawaii, I took a job with RCA, working 300 miles north of the Arctic Circle for the Defense Early Warning System. I had always wanted to go to the North Pole, but one year of isolation and adventure was enough for me, so I returned to Cleveland by way of Redondo Beach and a red Corvette, which I promptly crashed.

I met a girl in Cleveland, got married and started looking for a job. That change motivated me into a 32-year career on the railroad starting as an electrician and working up to foreman, and then supervisor, then division manager, regional manager, and, finally, director of labor relations. During all those years, I thought little about Vietnam. It became a footnote of my youth, a break in my Woodstock soundtrack, not something I talked about much. Then things changed. Kids started coming home from Iraq to hugs and praise and public demonstrations of patriotism.

Subconsciously, I started to feel resentful and angry. That is the way it should have been when we came home. But in the '60s and '70s, no one was able to separate the war from the soldier. I harbored a lot of resentment for a long, long time but was not particularly aware of how it affected me until I saw a short video about a young boy in a diner with his mother. A young soldier in uniform was also in the diner. The kid approached the soldier and said that his mother told him the soldier was a hero. Then he asked if he could shake a hero's hand. Of course, the young soldier shakes the kid's hand and says thanks. Meanwhile, the camera pans across the counter and focuses on an older man with a tattoo on his forearm; he is apparently a Vietnam veteran. He listens, and you can sense his resentment at the patriotic gratitude directed toward the younger soldier. Suddenly, the kid walks over to him and says "My mom says you are a hero, too," and he asks to shake the older man's hand. I cried, I was overwhelmed. I am tearing up as I write this. What the fuck were we forty years ago?

I stayed angry and silent for a long time until by accident I stumbled on the AVEL Vietnam website. That led to attending an AVEL reunion, which opened me to a brotherhood that I had missed for 40 years. I went to one of their reunions and immediately felt at home. My wife and I both felt we had discovered real people and felt entirely comfortable immediately. I started to acknowledge publicly that I was a Vietnam veteran, that I had served two tours and that I was part of a unique group from a tumultuous time in our history. I am not a nut about things, but I have started acting like a proud veteran and have started getting positive feedback.

Incredibly, less than two years later I was diagnosed with a type of cancer that the VA presumes is a result of exposure to Agent Orange. But it is not just cancer, I have other medical issues that all are listed by the VA as related to chemical exposure. I also experienced some 20 years ago a very serious auto-immune response following surgery for diverticulitis. I discovered while researching my cancer that the VA had been advised in a seminal study on the effects of Agent Orange that the immune system is "exquisitely sensitive" to Agent Orange. So, naturally, I now blame the use of chemical herbicides in Vietnam in 1967, '68, '69 and '70 for all the maladies I now suffer. Am I angry about that? Yes, I am! Am I angry about what my family has been put through? Yes, I am!

I'll file claims with the VA. I'll wear an orange wristband in honor of every Vietnam vet who has been killed or damaged by Agent Orange. I'll talk to people about it everywhere I go. I will never allow pity to enter into the equation. I will replace anger, to the best of my ability, with a sense of mission. That mission is to help bring Vietnam veterans out of the shadow of an unpopular war so we finally can be welcomed home and thanked for our service to our country.

The truth is I learned a lot about myself in Vietnam. I learned that I wanted to be good at everything I did. I learned that the way to succeed is to make sure that everyone around you is succeeding. I learned that it is much easier to misuse authority than it is to use it constructively; therefore, it should be exercised judiciously and questioned constantly. I learned that a little bit of insubordination can be a good thing. I learned that a good presentation can be as important as skill and knowledge. I learned that good communicators don't just speak well, they get people to listen and they become good listeners themselves.

In hindsight, I guess I was ready to be a soldier because I somehow believed that every man owes a turn in the military to his country if he expects that his country will defend him and his family should the need arise. I think the war I was asked to fight in did little to make America safer or to make the world safer. But that was not my decision to make at the time. I am especially angry that Richard Nixon extended the war in Vietnam solely for personal political gain, and that he did so at the cost of untold additional suffering. That is something I never knew about until the Lyndon Johnson tapes were unearthed. I am getting over my anger. It has taken a long time for me to become a proud veteran who can say I served two tours in Vietnam, got an Army Commendation Medal for service against a hostile foe, and got cancer from exposure to Agent Orange.

Specialist 5th Class Steve Nagle
3/17th Air Cavalry, 1st Aviation Brigade

I witnessed a fatal crash when two helicopters collided in midair. The smell of burned bodies and crispy meat haunt me as well. There is an overwhelming feeling, at times, of loss. Roadkill still bothers me today. Knowing how vulnerable I was flying at treetop level with no doors gives me a sense that every day is a good day. I find great satisfaction in helping other vets with their issues and struggles. We are all on this journey together. Some are doing better than others.

The first time I visited the Wall was very emotional for me. I left my family and approached the Wall by myself. Feelings overcame me, and I left without getting up close. The next day I returned and was able to find some names. The fact that it was constructed without any government money confirms to me that we Vietnam veterans are really our country's bastard children. I still don't feel support or thankfulness from our country. I guess it hit me hardest when the returning troops from Iraq got a ticker-tape parade in New York. "Where was my parade?" I asked myself.

How did the war change me? Well, I tend to be a positive guy. I enjoy each day. Mornings are the best. I read a story about a returning vet who went to the VA hospital to see a wounded comrade. The man in the hospital was not going to make it. "Please do something for me," he asked his buddy. "Each morning, welcome the new day. I won't be here to hug my kids. Hug yours for me. I won't be here to enjoy watermelon. Take a bite and savor it for me. When you love your wife, enjoy her for me as I will not be able to enjoy those hugs and kisses and intimacies." So, I live my life to the fullest for those who cannot.

I felt guilty for having a non-combatant role and guilty for having been to Vietnam.

1st Lieutenant Richard Kline
Special Services, 101st Airborne

I was intellectually against the war because I thought the Gulf of Tonkin Resolution was bullshit, and trying to fight a land war in Asia was really against conventional wisdom. I wasn't a protester, I wasn't marching in the streets and anything like that. I was living a very rarified, cocoon kind of existence in the drama class at Northwestern University. I went to graduate school to buy time so I wouldn't be drafted. I thought that by then the war will be over. Well, of course, it wasn't.

I was in rehearsal for my first actor's union job at the American Shakespeare Festival, Stratford, Connecticut, when I received notice to report for my induction physical. When I went to my induction physical I brought documents showing that I had a medical condition called phlebitis (blood clots in my leg) as a teenager, but at age 24 it was considered not medically exclusive. At my induction physical, there was one guy who showed up wearing a ballet tutu. These were strange times. After that induction physical I was classified as 1A. I started to freak out because I was against the war and in a university situation where everybody was pretty much liberal and not hawkish.

When my induction was approved, I appealed first at my local draft board in Long Island City, New York, and then to the head of Selective Service, New York. When that was denied, I considered moving to Canada or Sweden! Draft-avoiding friends suggested I shoot off my toe! I talked it over with my dad. He said, "Yeah, you can go to Sweden, Germany, or Canada but you will probably never be able to reenter the United States." That was the feeling at that time. He counseled me to make the best of it. So I said, "You know what? I'm just going to go in and make the best of it."

My feeling was pretty much against the war but I didn't want to go to jail, I didn't want to do what Cassius Clay (Muhammed Ali) did. I remember the day of my induction, they marched us into a room and had us stand there, four, five, or six abreast, and they said, "If you don't take one step forward, you will be against the Constitution of the United States and a felon." I stood there for two seconds and I went, "Pff, I'm stepping forward. I'm not going to jail." I just didn't want to go to jail, I didn't want to jeopardize my future as an actor and have a stain.

So, I was drafted into the Army during the height of the Tet Offensive in 1968. I went through basic at Fort Jackson, South Carolina, and AIT for combat engineers at Fort Leonard Wood, Missouri. I had been drafted for two years and signed up for an additional year for Officer Candidate School. After Engineer AIT I attended OCS at Fort Belvoir, Virgina, where I graduated as a 2nd Lieutenant in March 1969, sixth in my class. My MOS was 1330, Combat Engineer.

After I was commissioned a 2nd Lieutenant I was assigned to Fort Ord, California, where I became the Special Services Officer. I had a staff of 60 and a budget of nearly $1 million. I knew that eventually I would be getting orders for Vietnam, and when they arrived I was not surprised.

My trip over was typical for most soldiers. When I arrived, I was assigned to the 101st Airborne Division up at Camp Eagle near Hue in Quang Tri Provence. I was initially assigned to the 77th Engineers, where I spent my first two weeks. Then someone looked at my DD33, which was basically my military resume, and saw that I was in Special Services at Fort Ord. They were looking for somebody to be second in command of Special Services for the 101st. When the commander of Special Services left I was put in charge. At Special Services we booked bands, booked Bob Hope, and we had a movie circuit. We flew movies out to fire bases. I organized softball and basketball leagues and Olympic Games, which included ARVN soldiers from the nearby city of Hue.

We had about four or five rocket attacks while I was at Camp Eagle. We could hear them coming in but they were targeting the airfield. They weren't looking for personnel to kill per se and they never charged the perimeter while I was there.

I would occasionally hop a ride on a chopper with some 16 millimeter movies and fly out to some remote fire bases. One of my vivid memories was coming in for a landing at a fire base and all the trees were stripped bare. In the branches of these dead trees were all these little white parachutes from the flares. It was so bizarre. There was thick, green jungle and then, maybe 600, 700 yards around the fire base, all the trees were stripped. And because they were firing flares every night the white parachutes were caught in these dead trees. It was a weird sight, like Christmas trees.

I was very fortunate not to be assigned to the boonies. There were maybe two "life-changing" events that happened to me while I was in Vietnam. The first was that I was struck by lightning at Phu Bai airport, waiting to board a C-130 bound for Saigon. I was on a supply mission to get more sporting goods for our unit. I spent two weeks in a hospital in Cam Ranh Bay. When I got back to my unit, I was referred to as Lieutenant Flash!

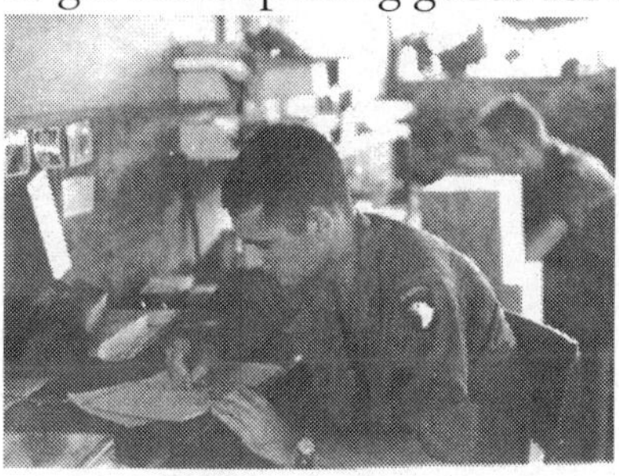

The second was a break-up of my engagement with my fiancée via MARS radio. To speak stateside, you had to go through a radio relay system with operators both in 'Nam and the US. The procedure was to speak into the phone and then say "over" when you wished to hear the other person speak. Well, some time had passed since communicating with the fiancée and when I finally did speak with her she basically said, "It's over—over!"

Coming Home

When I came home I processed through Fort Lewis, Washington. I'll never forget it—it was February 1971 and it was snowing. I was wearing short-sleeved khakis and I froze. When we arrived, they marched us to this cafeteria or mess hall and they gave us a steak dinner. The mood of the country was strained, in that anti-war protests dominated the news.

I was beset by what I call "double-guilt." I felt guilty for having a non-combatant role and guilty for having been to Vietnam.

I never ran into any protesters when I came home. But, it was kind of tough coming home. My friends were happy to see that I was home safe and they were just glad I was out of it. I think that was the prevailing attitude. I didn't get into any political arguments or moral arguments. It was mainly, "We are just glad you are safe. Now get on with your life."

Home

Once home, one of the first jobs I got as an actor was with a very prestigious company called the Lincoln Center Repertory Company, which was a year gig, four shows. There were all these actors who had been around, some older and some my age and I never opened my mouth about being a Vietnam veteran. I kept my service a secret. This was 1971, for God's sake, people were still marching in the streets. So, I just kept it under wraps. Obviously, my family knew and they were just happy to have me home. After a while, I became comfortable with telling people about my service. Some would ask, "Did you kill anybody?"

I didn't tell too many people about my service. It was when Vietnam vets were honored and appreciated that I gave myself permission to publicize the fact. I am a public figure, for better or for worse, and I finally put in my bio that I was a lieutenant and I served in Vietnam. I put the fact that I had served in the playbill of shows I was appearing in. I would get interviewed and it would come out in publications that I was a Vietnam vet and people would say, "Wow, I didn't know. Thank you for your service," and that really feels good when someone says, "Thank you for your service."

When I first started on the TV show "Three's Company," playing Larry Dallas in the late 1970s, I had the opportunity to appear on several game shows. On one particular game show we played for contributions to a favorite charity. When they asked me what charity I was playing for I responded, "The Vietnam Veterans of America." I had never contacted the VVA then. I gave them the address and told them to send any money I won to them. It was like 12 grand. Then I got all sorts of phone calls; "Richard, thank you so much." It is a wonderful organization.

I have been to the Wall a couple of times. It's very moving. While at OCS we had TAC Officers (Tactical Officers): they are the ones who scream in your ear and tell you to drop down and do 20, just like the DIs in basic training. When you get there they are like the devil and then when you graduate they shake your hand. Well, one of my TAC officers was killed in Vietnam. I looked him up on the Wall and you look at the name and then all the memories, good and bad, come back to you. I remembered him yelling at us and whatever, but it is sad. And the saddest part of all is, I mean, I don't have to tell you this, it was basically a bullshit war. I mean, it would be nice to go back in time and fight against Hitler; that would make some freaking sense. But this was so muddied and, God, I mean, the protests and all your pals you went to school with saying, "Oh, you're going over." It's kind of rough, it's kind of rough. Strange times, strange times.

Each time I went to the Wall was very emotional for me. Although I did not participate in combat, I felt strongly moved by the sacrifice of my fellow soldiers who died in a war that never should have happened. We stepped into a civil war, on foreign soil, that posed no threat to our existence thousands of miles away. Historians called it a "quagmire," and then came Iraq and Afghanistan.

I enjoyed leading the troops and becoming an officer. Yes, I was nervous about going to Vietnam but I didn't go AWOL. I went like you. You followed orders.

Staff Sergeant Frank Haught
D Troop, 3/5th Air Cav, U.S. Army

In 1969, I was living in Banning, California, with my brother, and I received the notice to get my physical in Los Angeles for the draft. When I took my physical, I looked around at all the hippies. I had never been to war but, after looking at them, I thought that it didn't look too good for our side. I returned home to Carmi, Illinois, and volunteered for the draft. Sixteen of us left Carmi on a train bound for St. Louis to be inducted. I was selected for the Army.

After basic training, I was sent to Fort Knox, Kentucky, to be trained as an 11D10 Infantry Scout. After our AIT my whole company was sent to Vietnam. I was lucky and got picked for the NCO Academy. But, they told us we would be sent to Vietnam anyway because of the shortage of NCO's over there. I figured if I was going, I wanted to have a little control over what I was going to be doing. And, sure enough, we got our orders: all of us were going to Vietnam. I was afraid to tell my mom that I was going because I knew how she was when my brother was over there.

I felt I was as well trained as I ever would be, and as a sergeant I wasn't supposed to be scared, but that changed. I received orders to go to the D/Trp. 3/5 Air Cav in Quang Tri, up in the DMZ as a door gunner. This is where I spent the next 12 months. I was discharged in September 1971.

Coming Home

When I left my unit, the guys in the platoon gave me a Cav. Scarf with my name and rank on it. As I got on the helicopter for Da Nang I was sad because I had lost some of the best brothers I would ever have, and then it started—survivor's guilt. As I got on the plane, I looked at all of these new guys coming toward me, and I was thinking, "How many of these men will never see their loved ones again?" I got on the plane and sat down. Everyone was quiet. After a few minutes, they came and took a couple of guys off the plane. A few minutes later a full bird Colonel sat in one seat of the now empty row where they were sitting. I thought, "You mean he couldn't sit beside an enlisted man so he can get home?" I was pissed, and so were most of the men, but still the plane was quiet. No one said anything. I wanted to scream with joy, "I'm going home," but when the plane lifted off and we got into the air the biggest cheer went up! We have left Hell! We're going home! As I sat there my mind kept going back to my unit. Why did I feel so bad about leaving them when I was going home? I felt I was betraying my brothers.

When we arrived in Fort Lewis, Washington, and got off the plane a couple of us kissed the pavement. We were back! No more killing or trying not to get killed and it is party time! A couple of us went right to the airport to go home, and as I'm walking through the airport, I see these signs protesting the war, people shouting at us, calling us Uncle Sam's baby killers, giving us the finger, spitting toward us, and yelling that we should have died over there.

I'm thinking, "This is not the country I was fighting for. Didn't they know we were fighting for their freedom?" I never thought I would see the day that kids our age would be protesting us. As I got home to the local airport people just looked at me. I was proud I did my duty, but I received no respect. Only stares and people shaking their heads. I thought I was happy but inside I felt betrayed and not welcomed at all. I felt I knew how the Confederate soldiers felt coming home.

I flew into Evansville, Indiana, and caught a ride home with a friend. I never told my mom I was coming home as I wanted to surprise her. When I got home, my mom came outside crying, and kissing and hugging me, and I thought she had aged twenty years. What has this war done to my mom? But, I was so happy to see her. I guess I didn't look much better. When my dad got home, he just smiled and said, "You made it," and never in his life did he ever say he was proud of me.

I tried to settle in. Carmi was the same, but I had changed. My friends were like high school kids, my girlfriend was in college and she was in the Hate the War movement, and I was a baby killer, so we parted ways. We didn't understand each other anymore. I felt old; she was young. I was a warrior, and she was a dumb college girl. I felt lost and out of place, so I started fighting in bars because I didn't like someone looking at me.

I was drinking and taking drugs. I didn't fit in anymore. I was out of place. The friends that were still around just wanted to know how many people I had killed. I felt like a hired gun with nowhere to go. My head was fucked up, so I started hanging around with other Vietnam vets who were just like me. I did drugs to escape my nightmares, staying out all night and sleeping during the day. This went on for months, finding no peace of mind; flashbacks and more flashbacks. It made me crazy: drinking, fighting, and just not giving a fuck. Uncle Sam sent us home without any help and nowhere to get it. My life was going down the toilet with no end in sight.

I needed more action. My life was boring, so I started doing dangerous things. I needed the rush like my Huey gave me. I still had my Hot Rod, but I needed more so I bought a Triumph motorcycle. All I wanted was the wind in my face like when I was flying. I needed the rush!

My friends finally started staying away from me. One night I was having a dream and jumped out of bed, smashed my fist through the wall, and broke a window. The next morning I knew, and my family knew, I was having problems adjusting. I had fought my demons long enough. I decided to enlist back into the Army to get my life adjusted back to where I felt normal.

Home

In 1975 when Saigon fell I was on my tank in Germany. My lieutenant walked up and told me Saigon had fallen. The North had won. A chill went through me; I had lost so many brothers over there including my best friend, Ray. All for nothing. All the emotional scars that I was carrying around. What that war had done to me and my family was all for nothing. I just walked over to the side of my tank and cried, all for nothing.

After all this I needed help, but I couldn't seek help because it would have been entered in my personnel file, so I sucked it up. I was discharged in 1977 as a staff sergeant E-6. My life was still in shambles. The Army didn't help me. Failed marriages and the war made me a bitter man. I was starting fights in bars again. I was a loner: I couldn't trust anyone. I couldn't get close to anyone until five years ago when I met my wife, Robin. Now I'm in group PTSD counseling and dealing with Agent Orange issues: Type 2 diabetes.

When I was in Vietnam, I prayed every time I stepped into a helicopter, and when I got back I thanked Him for another day. I don't think I would have made it if it wasn't for my prayers and my mother's prayers, so God was my savior and with His grace I am here.

I traveled to the Wall years ago. I went alone at night, and I made my peace with the brothers I lost over there. I had a talk with all of them and told them that they are not forgotten. I hope someday I can see them again, thank them again for what they gave for our country, and tell them how I miss them and how proud I am to have served with them, because they are all heroes.

I feel the war took my youth, made me into a man a lot faster than I wanted it. It made me a loner, bitter at this country for treating us 'Nam veterans the way we were treated. My mom and dad never thanked me for going to war. My dad never told me he was proud of me for what I did. I felt ashamed because my dad was not proud of me. I always felt going to war changed my life but not for the better. I think it made me a man that faced problems in my life. It made me grateful for all the simple things in life and knowing war is not a good thing. I'm proud of my flag and country and would do it again if asked.

Sergeant Tom Newton
A Company, 228th Aviation Bn., 1st Cavalry Division

Everything had changed. My friends had moved on. Many of them got married to get a deferment. I didn't see many of them, maybe one or two, and that was it. I felt out of place because a couple of them were smoking pot and taking drugs and I wouldn't go in that direction. They had long hair and were thinking differently than me.

My girlfriend was pregnant when I got home. I was totally upset; it wrecked me, it wrecked my homecoming, it wrecked everything. I didn't know what to do. I got through that on my own, no counseling, nothing like that. It ended in disaster, so I just went on to Fort Knox and that got me away from the neighborhood. I finished up my last five months at Fort Knox. So that was a nice welcome home for me. That didn't help either. From there I just continued on, doing the best I could on my own. I was finally discharged in August of '68.

When Saigon fell in '75, I kept saying, "All this for nothing." I did all of that for nothing and I was unhappy about it that we pulled out, and I just couldn't stand to watch it happen. I was just unhappy, and I began my lifetime of depression over it, wasting all that for two years, all that for nothing, and I couldn't believe it. I think we should have put up more of a fight, but I guess we chose not to, politically.

1st Lieutenant Charles Blake Clark
11th Infantry, 5th Infantry Division (Mechanized)

When I was a freshman in college I didn't do well my first semester and my mom said, "Son, if you don't do good you're going to flunk out of school and you'll get drafted and sent to Vietnam." I said, "Where the hell is that?" Well, I hadn't heard of Vietnam when I was a freshman. By the time I got drafted I knew where it was and I knew I didn't want to go there, but as it turns out I did.

I was drafted in 1968 but I was able to postpone my induction in order to graduate from college. I was finally inducted on January 2, 1969, and proceeded to Fort Dix for basic training. After I completed basic I was sent to Fort Leonard Wood, Missouri, for individual training. I qualified for Officer Candidate School so I applied for that and was accepted. I was sent to Fort Benning for my OCS training. My intention was to try and stay in the States as long as possible, hoping the war would end before I had to go. That was really naïve, but that's what my plan was. It didn't work. I graduated from OCS with an infantry MOS and then I was sent back to Fort Dix for ten months before being sent to Vietnam.

I left for Vietnam out of Travis Air Force Base near San Francisco. It's funny; I went back to Fairfield, California, a couple of times to perform at a comedy club there, and the first time that I went I thought, "This place looks familiar," and then I saw Travis Air Force Base and I told my wife, "This is where I left to go to Vietnam."

It was September 1970 when I went over. I flew into Tan Son Nhut and caught a C-130 up to Quang Tri. My first duty station was as a platoon leader with Alpha Company, 1st of the 11th Infantry, Fifth Infantry Division. Our unit was actually a brigade and I think it consisted of one airmobile infantry battalion, which was mine, one mechanized infantry battalion, and one cavalry battalion. We were up in Northern I Corps.

It got rough at times but for the most part there wasn't much to do. We'd go out after the enemy and chase them and they'd see us and chase us. In my platoon no one was really gung-ho. Everybody kind of had the attitude, "I don't want to be the last guy to die in Vietnam." They were aggressive when the time came but it wasn't, "Okay, today we're going to go out and find some enemy and kill them." When the monsoon set in they would send us out on patrol but the North Vietnamese were smart enough to stay inside.

Normally we went out in the jungle and disappeared. If we had contact it was by accident. The enemy didn't like it because if they knew where everybody was then they could plan accordingly and do what they wanted to. But when you went into the jungle and did the same thing that they did, they would start avoiding the area. It was terrain denial and we were really good at that.

I never asked my guys to do anything that I didn't do. A lot of the guys that I went to OCS with became officers like that. We had our men's welfare in mind over everything else, even the mission sometimes. And there were times when I just refused an order because it was tactically questionable, and I felt it would endanger my troops' lives for questionable results so I just said, "No." And they said, "You are refusing an order?" and I said, "Yes, pretty much. What are you going to do?" I just called their bluff. "Come here and get me, come on. There is a chopper out here, I will get on it and take my ass back. Come on, do it." Of course, they wouldn't, because they didn't have any more lieutenants. When my guys saw me standing up for them, they stood up for me.

I stayed with my platoon until March 1971. I was then assigned to Headquarters Company of the First Battalion, 11th infantry. I became the support platoon leader and I didn't have to go out in the bush anymore. My main job was to be the officer in charge of convoys. I was also responsible for getting aircraft to resupply the troops in the field. Anyway, I would start taking out these convoys, and there were only two roads in the area so the enemy pretty much knew where we were going to be. I don't know about after I left but the two guys before me that had the job were both severely wounded. I went out on the first convoy and we were shot at. When we got back I said, "Hey, can I go back to my platoon? Go back in the bush?" And they said "No, this is what you need to do." So that was my job for the remainder of the time that I was there. It was pretty intense.

Coming Home

I came home in August 1971, so I served 11 months there. Our unit stood down and we left out of Cam Ranh Bay, flew to Japan, where we stayed for a few hours, and then flew on to Fort Lewis for processing out of the Army.

After we were released we proceeded to the airport at Seattle-Tacoma, where there were people making snide remarks to us. I remember I was with one guy who was a second platoon leader, and I was telling him goodbye. I remember just standing there saying, "I don't know what to say, I know what we have been through together, I just don't know what to say." People were looking at us and saying "blah blah blah," and I said "Hey, fuck you!" and my buddy said, "Hey, forget them, forget them, it's not important." We hugged and I was crying. He basically saved my platoon during Lam Son 719. I boarded my flight and started my trip home. I had to go through Chicago and then to Atlanta. I was walking through the Atlanta airport and I have my uniform on with all my attaboy badges and ribbons, and this fucking slimy, dirty guy calls me a warmonger and murderer and I just yelled, "What the fuck do you know?" and just kept walking. I came home, took off my uniform and I didn't talk about it for a long time. I put the uniform in a drawer and left it there. My mother actually wrapped it up and put it away. She wrote on it; "The uniform Blake wore home from the war in Vietnam." When she died I found it amongst all her stuff. I still have it somewhere. It wouldn't fit now. I weighed about 135 pounds when I came home from Vietnam. I was so tired of eating C-Rations, I couldn't stomach them anymore.

Home

After I was home I had to report to the local Army Reserve unit, so I went there and even though I was in civilian clothes I figured I'd better do this militarily. So I walked in to see the commander, stood at attention in front of his desk and said, "Sir, Lieutenant Clark reports." And he said "Wow, this is great, are we happy to see you. An Infantry lieutenant with medals and combat experience, oh boy. We have a good time here, we have meetings once a month, we do two weeks of summer camp." And I'm thinking, "Oh, fuck." And he just busted out laughing and then he said "No, we do this to all you guys. Don't worry about it, you have done your time, you have served your country, get out of here." I went, "Thank you, Sir" saluted, got the hell out of there, and never went back.

Things were different when I got home, totally different. My friends didn't realize what I had been through and they didn't want to hear about it, so I just kept it to myself like most guys. I had trouble dating because girls didn't want to go out with Vietnam vets and my short hair was a dead giveaway.

After I was home a while I got a job at the Georgia Department of Labor employment office. It was to help veterans get re-assimilated into society and help them get a job. My job was to go around and try to get people to hire veterans. If companies would pay them at that time I think it was 3.25 an hour, the government would match it and the vets could be trained at the job. I remember being in a mechanic's shop, it was also an auto parts place, and the owner said "Well, $6.50 an hour is a whole lot to pay a drug addict," and I said, "What?" "All them Vietnam guys, they are all drug addicts, murderers, or cowards." I yelled, "Hey I'm a Vietnam veteran, God dammit." And I grabbed a rubber hammer and I said, "I will show you what a fucking coward is!" The next day the head of the department told me, "We are going to leave you in here to talk to veterans instead of going out there and trying to get people to hire them." So what happened was these guys would come in and they would send them to me. I would ask them what unit they were in and all that stuff, and then I would tell them that they didn't really have a job right now for them, but here is this guy's number, he was in the same unit, why don't you guys talk to each other. Actually, by accident, I kind of started the first vet-to-vet counselling thing. I remember this guy came in who was with 101st and having trouble dealing with things. I knew a guy who was also with the 101st, in the same battalion but a different company. So I said, "Here, just tell him I gave you his number. If you feel you need to talk, call him."

When the program ended I had to go find another job. I tried a bunch of different things and nothing worked out. I even tried to own my own business and that didn't work. So, I moved to California in 1980 to do comedy, and I was floundering, I didn't really have direction. And then one night at the comedy store Richard Pryor was there talking to a bunch of us comedians. He said, "If you can write comedy about things that bother you and make people laugh about the things that used to make you cry, then you have a lot of material, but also it will be very personal. Nobody else can steal the jokes from you because they are about you." None of the others took it to heart, but I did. I thought about it and said to myself, "Wait a minute, what's different about me? I was in combat, I went to Vietnam." These other guys were all about ten years younger than me and none of them had even been in the service. So I started talking about that, and one thing led to another and I got on "The Tonight Show" and Johnny liked me, so I was kind of up and running and I did pretty good.

By the time I started dating my wife, my hair was longer, moustache was thicker, and I had sideburns. I didn't say anything about Vietnam.

One night we were sitting outside and she said, "You have a lot of unresolved anger." And I thought, "No, I don't. Bullshit." Actually I still have a lot of unresolved anger. But I said, "No," and then I said, "I'd better tell you, you'll find out sooner or later. Right before I met you I was in Vietnam." And I thought she was going to say, "All right, that's it, we are done, get the fuck out of here." But she said, "Well, that explains a lot," and so the next week I asked her to marry me and we have been married for 44 years.

I did have some anger issues when I returned from Vietnam but I never went to the VA or had any therapy. I kind of worked out my demons by being married to my wife and doing comedy. Doing comedy and making fun of it is the way all comedians deal with adversity. That's how we become comedians. So I started talking about Vietnam on the stages of America and people accepted me and laughed at my jokes about being a soldier and a veteran. People would come up to me to talk and I realized I wasn't the only one feeling this way. And then I watched some of these shows on the History Channel, and they interview veterans from World War II, Korea, Vietnam, and now Afghanistan, and I realized that I have posttraumatic stress. I don't need to go into therapy to find that out. I realized that everybody has the same thing to a degree that I do; I am not different. I never sought any treatment, because I accepted it. Also, I was older than most of the kids over there; I had already graduated college. There were a lot of college grads in my platoon. I think I had eight or nine in my platoon, which is a lot.

I never took combat personally. My attitude was, "Yeah, why shouldn't they try to kill me? I am in their country, I am a soldier, I am trying to kill them." So I never took it personally. They aren't trying to kill me, Blake Clark, they're trying to kill Lieutenant Clark.

I haven't been to any reunions but I have done a lot of shows for veteran groups. I would always do USO shows because the USO was the only friend that the soldier had, so I always do whatever I can do for them. I have been to Afghanistan and Iraq twice. Once with Jim Carey and a bunch of other comedians, and once with Robin Williams, bless his soul. I flew with him to Washington, DC, and then we went over with the Chairman of the Joint Chiefs. We went to a whole bunch of different places. We went to Iraq, Kuwait, UAE and Qatar.

I have been to the Wall. I went there alone. I was doing a gig for the veterans in that area, in Alexandria, Virginia, and went to it at two o'clock in the morning. I called my wife, bawling, and she said, "I'm sorry I'm not there with you." I said, "No, you know I came home alone, I might as well come here alone." So that was my trip to the Wall. Pretty emotional.

We turned everything over to the ARVN's and we all just knew that this wasn't going to last. I came home in August 1971, and in May 1972 Quang Tri was overtaken by the North Vietnamese. Quang Tri city was levelled; I mean, literally, not one stone left, it was gone. And I was sitting there watching the newscast and I am thinking, "What the hell did we even go there for? What the fuck did we do? What did we accomplish? What did all those guys die for? What did all those people get wounded for? What? What? What did we do? Nothing."

I would never consider going back to Vietnam. No, I don't want to go back. It's not because I am afraid I will have bad memories, I am afraid I will go there and see a Motel 6 on Hill 240 and I just can't deal with that.

Looking back, I think there is a saying that is, "A million-dollar experience I wouldn't give you two cents for." I wouldn't take 10 million dollars for it but I wouldn't do it again for anything. But I think it made me a better person. I have something by which to measure the rest of my life; nothing that I have lived through yet, knock on wood, has been as bad. And I can't envision anything other than the loss of my sons and my wife that would be as bad as that was. One thing that I tell people that is hard to explain is, this is kind of like being old, being 70 years old; it's kind of like being in Vietnam, you feel shitty all the time, you hurt, you ache, your friends are dying and you know your time is coming. The only difference is that . . . the difference between old age and war is that war is just for a year and there was a chance you would get out, you would survive it, which I did. This is for the rest of my life, I am not getting out of it alive. Made me able to deal with people dying, you did that at an early age.

Captain Forrest (Woody) G. Hester III

2nd Bn., 12th Artillery, 23rd Arty. Grp. U.S. Army

I enlisted in 1966 to avoid the draft. I could see that it was inevitable, and I wanted to try to take advantage of the Army's great training programs to learn a useful skill. A good Army recruiter helped me do that.

After basic training at Fort Leonard Wood, Missouri, I was trained in basic electronics, then went to a year-long school at Fort Bliss, Texas, and became a computer and radar repair technician for the Nike Hercules Missile System. I graduated a Spec. 4 and served in my new capacity at Fort Bliss for exactly two weeks. A friend of mine named Frank and I had decided to take the test for OCS, and we were both accepted. We graduated from Field Artillery OCS at Fort Sill in August 1968. Almost the entire class went to Field Artillery Officer basic two blocks down the street, then straight to Vietnam. Frank and I, because of our Air Defense Artillery training, went back to Fort Bliss for Air Defense Artillery Officer basic. After that, I was assigned as the assistant range control officer of McGregor Range just north of Fort Bliss. Frank was assigned to the Radio Controlled Aerial Target Company at McGregor. During this time I went home on leave, married my high school sweetheart, and brought her back to Fort Bliss. My friend Frank met, fell in love with, and married a girl from Mexico who was attending University of Texas at El Paso on a student visa. Our wives became good friends, and we led a nearly idyllic life at Fort Bliss.

While I always knew I might be sent to Vietnam, I had grown to believe I wouldn't. I was next in line to become the range control officer at McGregor and I thought maybe the Army would allow me to serve out my obligation there. The life my wife and I had together at Fort Bliss, with good friends to share it with, and a good income, was a dream come true.

One morning, the range commander, Colonel Weafer, called me and asked me to come see him. I reported as directed and he quietly handed me orders to Vietnam, then waited until I had read them. In that instant, my idyllic life came crashing down. A million questions raced through my mind. Things like: What will happen to my wife? Where will she go? How will I tell her? When I looked up at Colonel Weafer, he said, "I have to give you these today, but I want you to know that I'm working through channels to get them rescinded. I need you here, and can't let you go." I argued with myself the rest of the day about whether I should even tell my wife but, at the end of the day, Frank called me. He had received the same orders so I decided to tell my wife, and everyone, my circumstance. My wife was scared. By that time, three of my classmates from OCS had already died in Vietnam. Frank's wife Olivia was scared too.

Colonel Weafer called me a few days later and told me to come see him again. I reported immediately, and he told me I was not going to Vietnam after all. I asked about Frank. He said that Frank was going. Then he told me that he had had to cut a deal with the Department of the Army and that he was sending another second lieutenant from the command in my place. It took about two seconds for that to sink in and then all the emotion of the past several days came bubbling out. I politely refused the deferral. I told Colonel Weafer that I would not permit that to happen. I told him that I was going in my place, period. Colonel Weafer kindly but very firmly said: "I understand how you feel. This is the Army, and the Army comes first. What you want, and what I want, isn't important. We have to do what is best for the Army. Now, go back to work."

I asked my driver to return to Range Command without me, and I walked the half-mile or so just to get my head cleared before talking with others, but, I could not get it cleared. I felt like I was part of something that felt like cheating, and the fact that someone would go, and maybe even die, in my place was too much pain to bear. I wound up going home early that day and spilled my guts to my wife, Frank, and Olivia. My wife was thrilled that I wasn't going, but had the same feelings I did about someone going in my place. Frank, great friend that he was, talked it through with me for a long time. He helped me sort out all my feelings and I finally accepted the fact that I had no control over what was happening, but I didn't like it.

Two weeks later I received another call from Colonel Weafer. I reported to his office, and he wasn't smiling. I thought I was in trouble. He handed me my orders and said: "Looks like you're going to Vietnam after all, and there's nothing I can do this time. Several others from the command got orders too." As I absorbed this new information, I was numb with conflicting emotions. On the one hand, I felt vindicated: on the other hand, I was again worried about how my wife would take the news, and I was worried about where she would be while I was in Vietnam. When I broke the news to her, she wasn't surprised. We lived in an Army housing area, and everyone else was going, too. We accepted it and began planning for a major life change.

My wife Lou wound up going home to live with her folks in Illinois. Frank had not left yet, so he and I raced down to records and made arrangements for each of us to escort the other home should he die in Vietnam. We figured that would get the survivor out of the war for several days, and one of us would be there for our wives if that terrible day came. Frank left for Vietnam before me, and I followed after a one-month leave. We all tried to put the best face on it, but it was a sad time. The night before I flew to Oakland, Lou and I didn't sleep at all. We just clung to each other through the night. We had been sweethearts since I was 15 and she 14. We could not imagine being apart for a year, and, if the unthinkable happened, forever. Privately, I was worried. The news from Vietnam was bad, and there were often tributes to fallen soldiers and obituaries in our local paper. I knew full well that I might not survive. After a stoic farewell at the airport, I made my way to Oakland, processed, and was assigned a flight to Ben Hoa AFB via Yokota AFB, Japan, the next day. I spent my last night on US soil at the Oakland Army Depot Officers Club, where I ate dinner and ended up at the bar with a bunch of other guys headed for Vietnam, too. One guy named Andy, who I talked to the most, shook my hand and gave me a hug as we parted company that night. He had tears in his eyes, and that shook me up. I had only known him for two hours. The next afternoon, as we all stood in line for our flight with our personnel files under our arms, Andy butted into line right behind me and said, "Hey. How'd you sleep?" "Just fine," I said, "I was really tired." Then he said, "We're gonna be just fine, buddy," and I believed him. I don't know why, but I did. To this day, I'd love to find Andy. I never knew his last name, or where he was from. We parted company at Bien Hoa AFB. We really connected in a short time, and his assurance as we boarded the aircraft carried me a long way. It meant a lot. I hope he made it home.

The year in Vietnam changed my life forever. For three months, on alternate days, I flew in the back seat of an O1E Birddog (L-19) as an air observer. On my off days, I served as the battalion's S-5 (responsible for indigenous hires and relations with the local Vietnamese). Then, for three months I served as the Service Battery XO & ammo officer. Finally, for the last six months I served as the Service Battery commander. While my tour was quieter than many describe, it had its moments. I could have died, but didn't, at least three times. My plane took rounds once, I was nearly crushed between a Conex and a CH-47 Chinook Helicopter while airlifting supplies into Cambodia, and I was pinned down all night in a firefight between ARVN & VC right outside of the gate to Tay Ninh. Eight men in my battalion did lose their lives: one in my Service Battery, when I was CO. Another member of my service battery when I was XO went missing and was never found. All these things, along with the constant mortar attacks, seeing death first-hand (on both sides), living in the dirt, dust, mud, and heat of Vietnam, had a profound effect on me. It gave me a reverence for life I had never known. I promised myself that, if I returned to "The World" I would be different. I felt a new purpose for living I had never felt before. I aspired to do things I had never aspired to before. Instead of allowing life to happen to me, I vowed that I would create a life for myself and my family that was centered around family, friends, community, and service.

My wartime experiences changed me profoundly. Four months into my tour, we got a new battalion commander. His name was Al Murphy, Lieutenant Colonel. He was, without a doubt, the best leader of people I have ever been exposed to. Working with him was a privilege. Learning from him was a gift. Because of him, I re-enlisted for an additional two years in the Army and extended my tour in Vietnam, so I could stay with him until he left. He helped me learn things about myself I didn't know. He helped me learn that I could do things I didn't know I could. He awakened the leader inside of me and showed me how to do it right. It was under him that I served as the CO of our Service Battery at a very busy time. That included the invasion of Cambodia when our three firing batteries were moved so far inside Cambodia that we had to resort to unconventional methods of re-supply. For me, it was an incredible time. Ultimately, when I left the Army, I carried all that I had learned from that experience and Al Murphy into my civilian world. I firmly believe that all the success I have had in organizational life is a direct result of the time I spent in that crucible of learning with Al Murphy.

Coming Home

Colonel Murphy left about a month before I did. He and I had become close friends in addition to close working partners. The balance of my tour was so short, I had a hard time connecting with and finding my focus with a new battalion CO. I spent most of my remaining days orienting my selected successor. The time went fast and, before long, we held the change of command ceremony (for Service Battery). I climbed in the jeep, and my trusted driver/RTO drove me to Group HQ to out process. When we arrived, he gave me a big bear hug and a unique plaque he had had made in the village outside of Phu Loi. It had our unit crest on it, an outline of South Vietnam, a South Vietnamese flag, and an inscription. Sergeant Johnson had been my driver/RTO/friend for the entire time that I was Service Battery CO. He shielded my body with his own the night we were pinned down outside of Tay Ninh. As he hugged me, I was overcome with emotion. All I could do was hug him back, and the next thing I knew I was watching him drive off down the road. I had a lump in my throat as big as a basketball and, while I wasn't sobbing, the tears kept leaking out of my eyes. It took me several minutes to regain enough composure to report to the processing center. I don't know if I ever knew his first name. I know he was from Mississippi, and his last name was Johnson. I really want to find him. I hope he made it home.

We were warned on the plane to change into civilian clothes at the first opportunity. We were told that many soldiers, in uniform, were being harassed on the streets and in airports. Because my wife had flown out to California to meet me, I took their advice and others on the plane did the same. My wife Lou was there when I walked into the terminal. My old friend Frank had just been discharged from the Army and actually lived in Oakland. He was standing right next to her. His wife Olivia was back at their house preparing a meal for the four of us. As soon as Frank saw me, he gave me a quick thumbs-up and a big smile. I can still remember the moment as if it all happened yesterday. We held each other for a long time. All of the other guys had to walk around us. In short order, the three of us ducked out of the airport and went straight to Frank and Olivia's apartment. Once again, just as it had been 15 months before, we were four friends together, and we were so happy. Olivia and Frank had a child while Frank was in Vietnam (little Olivia), and getting to meet her was so special. We stayed up way too late catching up. It was so good.

Lou and I flew from California to Springfield, Illinois, which was our hometown. My parents, her parents, and extended family from both sides were there to greet us, maybe 30 people. It was very nice. There were tears, flowers, and hugs all around. Among the crowd was my Uncle Jack, who had served in Korea (Purple Heart) and was a very proud veteran. He gave me a big hug, but then said, "Why the hell aren't you in uniform?" I was immediately embarrassed. To this day, I regret taking the advice on the plane. I should have worn my uniform all the way home.

After getting home, I sometimes felt like I needed to pinch myself. I did silly things like stand in the bathroom admiring the plumbing. I took long showers. I lingered over my meals and stared at the American faces in crowds. I was apprehensive driving in the dark. Loud noises startled me, and I sometimes woke up in a sweat, but that all subsided in a relatively short time.

While on leave at home, I was treated well by friends. Because I had been on television news and in the newspaper when I was awarded a Bronze Star, most in the community knew I had been in Vietnam. Some were silent, but most were friendly and positive. One person asked me if I was bothered by anything I had had to do. I just said; "Sure. Normal people are bothered by wars, and I'm normal. But they were all things I had to do."

Home

To this day, I am amazed at the number of people I hear who seem to be experts about the Vietnam War. The revisionist history about that event, throughout all levels of society, is astounding. The people I felt disconnected from the most are those who think they understand that war, but don't.

When Saigon fell I was finishing my first degree at Ferris State University. I was traumatized by the events unfolding on the nightly news. When I was serving as our battalion S-5 in Vietnam, one of the things we did was connect with and support an orphanage in Phu Coung Province. When the news about the crash of an Air Force Transport carrying Vietnamese orphans out of Ben Hoa reached the United States, I was devastated. I could think of nothing else for days. I attended class but was only present. I just couldn't get over the profound grief I felt for those kids and my anger at all of the foolish politics that lost a war for us that we were winning when I was there. I'm still angry, and my heart still aches for those kids.

Compared to many, I didn't have serious problems afterward. I drank too much for a while but got past that pretty quickly. I had so much resolve to build an incredible life for my family that it consumed me. I focused on school, work, and my family.

For me, my experience in Vietnam was a formative time. It transformed me in very positive ways. I came back a much better man than the boy who went. As I had promised myself, I plunged headlong into making a good life for my family. I never joined the VFW, American Legion, or AMVETS. I put my uniforms away and have never worn them again. I donned civilian clothes, got a clean haircut, went to college, then to work, then to career- and family-building. My life and the lives of my wife, four children, and ten grandchildren have been blessed beyond measure, and, by the way, the G.I. Bill paid for most of my college education. I'm convinced that if I had not served, my life would not have been nearly as much.

Major General James Mukoyama
Company B, 4th Battalion, 39th Infantry, 9th Division
MACV, II Corps G3 Adviser, U.S. Army

After high school I went to the University of Illinois at Navy Pier for two years and then downstate at Champaign. That was 1963. I was commissioned in '65 as a regular Army infantry officer. For me to do that, I had to get two waivers. Number one, I was too short. There was a height minimum of 5' 5" and I was 5' 4½". I also had to get a waiver because I had a tumor on my temple. I had an operation, but they were concerned that it would preclude me from serving the full 20, 30 years.

After I had been commissioned, I took a year of leave. It was called excess leave in the military, because I wasn't paid during that time, but I was able to go to school. And so I got my master's degree. Now I was a more valued asset because I was coming in with a graduate degree. So, in 1966, I went to the infantry officer basic course and then I went to airborne school. After that, I volunteered for Vietnam because I was an infantry guy.

But the Army sent me to the DMZ in Korea. It turned out to be a very good assignment, because, on the DMZ, we were doing combat patrolling, every day. The 13 months I was there, which was November '66 to December '67, my battalion sustained 12 KIA and about 40 wounded. No one knows about that. In those days, we might be losing 100 guys a day in Vietnam, so Korean casualties were not big news.

In Korea, I had my first experience with mass casualties. We had a barracks blown up by an enemy sapper attack. We had three guys killed and about 35 wounded.

When my tour in Korea was over, I volunteered for Fort Campbell or Fort Benning because I had not been in an airborne unit and I was airborne qualified. Of course, they sent me to Fort Lewis, Washington, as an AIT Company Commander. I was a captain by then. Have you ever heard of a guy by the name of David Hackworth, Colonel Hackworth? I affectionately refer to him as "Hack." But in those days, he was a lieutenant colonel, and I was a captain. He couldn't pronounce "Mukoyama" so he called me "Mook," and I would call him "Sir." He was a legend. Serving under him was a great honor until he was sent back to Vietnam.

Everything I had ever asked for from the Army, I never got. Finally, I said, "Screw this. I'm moving on with life." Then Hackworth sent me a letter from Vietnam. And he said, "Mook, what are you doing at Fort Lewis? We got a war going on. If you want a company, it's yours."

It's like I hit the lotto. So, I called the infantry management office at the Pentagon, and I said, "Hey, look me up. You'll see I'm an infantry captain, and I've never been to Vietnam." I volunteered for Vietnam, again, but with the stipulation that I be assigned to the 9th Infantry Division, because that was where Hackworth was. They agreed, and scheduled me to go in April, which gave me time to get things in order. Everybody in the world told me when I hit Tan Son Nhut that all bets were off. They needed infantry captains everywhere and we were a dime a dozen.

When I got there, Hackworth had all the skids greased. They immediately put me on a helicopter and I flew down to Dong Tam, which was the 9th Division headquarters. They had a five-day orientation course for newbies down in the Delta, the basic "How to Stay Alive" course. So, I was there for a day and a half, and I get pulled out and put on a helicopter. And there I go, down to our fire support base.

I got off that helicopter, and I was madder than a wet hen! I saw Colonel Hackworth and I said, "Sir, how can you do that to me? I'm a new guy; this is a five-day thing to learn how to stay alive here, and you pull me out?" And he said, "Ah, Mook, don't worry about it. I know your ability. I need you here."

From that point forward, I always knew that God was with me. When I was in Vietnam, I never once worried about my personal safety. It never entered my mind, because I always knew if God was going to take me, I'd be in a better place. That permitted me to put all my time and effort into my job commanding soldiers. That was what kept me awake at night: trying to anticipate what the Vietcong was going to do and what would I do. As I've said many times to people who have not been in the military: if you're in business, and you go bankrupt, you can rebuild the business. But if you're leading soldiers in combat and you screw up, it costs lives, the most precious commodity that our nation has: the great young men and women who I, as a commander, was entrusted with their safety. So, my faith has always been extremely important. And it still is, to this day.

One day we had just had a battle with the Vietcong, and my company had overrun their position. We had killed numerous Vietcong, and there were three dead bodies at my feet. I'm the company commander, and the unit is most vulnerable right after a victory. It's just human nature to let your guard down. Well, I'm the professional, I'm the guy in charge. I know this.

So, I'm on my radio, talking to my platoon leaders, kicking rear ends and taking names, and saying, "Hey, listen. You have to re-organize your units, re-distribute ammunition, take care of your wounded, look for enemy avenues of approach." And while all this is going on and I'm on the radio, I stopped. And I looked at the bodies at my feet, and I realized that something had happened to me. Something had hardened my heart: where only moments earlier, these were alive human beings, actually, children of God who had families. They had emotions, loved ones. And they were fighting for something as important as I was and I was in their back yard. And yet, I was just treating them as if they were bumps on a log. And then I thought about Jesus' Sermon on the Mount, where he told us to pray for our enemies. So, in the middle of all this stuff going on, I said a prayer for the three Vietcong and their families. Now, I didn't get on my hands and knees, have a big ceremony, and all that stuff. This all took half a minute, 45 seconds. But it's something that always stayed with me for the rest of my life.

After this, I had a little more compassion for them. But I think it was more respect than compassion because, as you know, for soldiers to be able to go out and kill people, you almost have to dehumanize the enemy. So, what do we do? We call them "gooks," we call them whatever names we can call them, to attribute something less than human to them. But it increased my respect for them.

Our unit was to be withdrawn from Vietnam and Hackworth wanted to stay in-country, so he went up to Pleiku as the senior G3 Advisor for II Corps Headquarters. So, when the orders came for us to rotate out, I called him, and I said, "Sir, can you use me up there?" The next day, I had orders.

So, I went up to Pleiku, and I was with Hackworth for the rest of my tour. I was the G3 Plans Advisor of the II Corps. It was like the Pentagon. It wasn't really that dangerous: now and then, we would go out into the field. I had an incident that was a close call. But it wasn't a big deal. So, towards the end of my tour, I submitted my resignation, to be effective when I got back.

Coming Home

I came home in April 1970 on a commercial plane. We flew into Travis Air Force Base. I kissed the ground when we landed in California. Then, we were bused to Oakland. I don't remember any protesters although I do remember them telling us not to wear our uniforms in public. I was an infantry officer; I wore my uniform. I just thought, "Hey, what is this?"

I lived a Norman Rockwell life. My church was extremely supportive of me when I was in Vietnam. I attribute that to a lot of things, the fact that I had been so active in my church. But also, we had a lot of vets—World War II veterans—in our church. And these guys were my heroes.

My church sent me packages and letters. But, most importantly, I knew people were praying for me. And I don't care what anybody says, the power of prayer and knowing that people were praying for you helps. It really makes a difference.

My return was a lot different than a lot of my peers. My church welcomed me back like I was a hero, especially the World War II vets. And then, they wanted to bond with me, find out what my experience was, and they were very proud of me. I was an infantry guy, and most of those guys were infantry guys who fought in World War II. My family supported me like crazy. My mom and dad had a map of Vietnam in our kitchen. And they plotted out wherever I was.

I never told my mom I had a Purple Heart. You know how it is: you pack up your stuff, and you ship it home. Well, I had a box I just threw all my stuff in. In the box was my Purple Heart. So, it gets home, and my mom opens it up. That wasn't real good.

I wanted to stay in the Army at least 20 years. At that time, the Reserves and the National Guard didn't have a lot of combat-experienced guys. I felt that I had something to contribute, so I signed up for the Reserves. I was very blessed. Just like in Vietnam and Fort Lewis, I had great NCOs who made me look good. And I had commanders who didn't cut my head off when I made a mistake. I signed up with the 85th Training Division, in the Chicago area, which was an infantry training division.

I commanded at every level within that division, up to assistant division commander. So, I was a company commander, I was a battalion commander, I was a brigade commander. I was in that division for 18 years, so I knew that division and the soldiers knew me. And it was really a very fine experience.

Home

Four years ago, I had a heart attack. Now, up to that point, I had been in very good shape, physically. I did have a slight problem with my kidneys. They started to deteriorate. It wasn't to the point where I had to do dialysis. But my nephrologist said, "You have to watch this." But then, four years ago, I had a heart attack. And it wasn't that big of a deal.

I got up in the morning one day and I felt this pain. I was getting ready to go to work, and I felt this pain in my chest. But it wasn't a severe, pounding pain. I wasn't sweating, I wasn't nauseous, I didn't have a temperature, fever, or anything. There were no other symptoms that might lead me to believe I had a heart attack. I thought it was something I ate. Well, it didn't go away, so I told my wife, "This is not good."

My wife drove me to Glenbrook Hospital. I walked into the emergency room and said, "Hey, I have this pain in my chest. I don't know what it is." So, they do an EKG, and they said, "You had a heart attack." So, suddenly, doctors and nurses come swooping into the ER, and I'm being prepped for the operating room. I'm lying on this gurney. And I say my standard mantra; "Every day is a great day. I have my faith, I have my family, and I live in the finest country in the world." Well, you know, that comes from my combat experience. There were times I didn't know if I was going to be alive the next moment, much less see the next sunrise. Well, when you're in that position, what's important in life became clearly focused. And to me, it was faith, family, and living in this country.

So, here I am on a gurney, going into the operating room. And I say to myself, "Self, can you say today's a great day?" And I said, "I absolutely can, because 45 years of borrowed time from Vietnam, comrades that we had that never made it back, have never had the honor and privilege to be married, to have kids, to see the internet." But most importantly, I had my faith. So, in 24 hours, I had three operations.

They did an angiogram, and they found out that my LAD—main descending artery—was 90% blocked. So, they did an angioplasty and put a stent in. By ten o'clock in the morning I'm in the recovery room, thinking, "Well, okay, that's the end of it." Well, it wasn't the end of it. What happened was, my heart cavity started to fill up with liquid. So, I had to go back in, and they had to put a hole in my chest so they could connect a tube to it to drain this liquid, which took about 24 hours. The third thing that happened was, my kidneys failed. So, I had to go back in, again. They had to put another hole in my chest so they could hook up the dialysis machine to me.

So, I'm on the table. The first operation, they asked me my name and date of birth: "Jim Mukoyama, August 3, '44. Every day is a great day. I have my faith, my family, we live in the finest country in the world." I can't tell you the effect that that had on the doctors and the nurses. The first operation, the doctor said, "What is your faith?" And I said, "Well, since you asked me, I'm a Christian. Christ is my savior." I said, "You are a skilled physician, skilled nurses. But I'm in God's hands, and whatever He decides, I'm okay with. So, let's get on with it."

I was on dialysis for three months, then I got a donor. The donor was our daughter, but our daughter is not our biological daughter. We adopted her 35 years ago. And she was a match. God knew, 35 years ago, that I was going to need that. Another thing is that you can't ask your kid to donate an organ, right? She volunteered, and her husband volunteered. And they tested her first, and she was a match. So, they said, "That's it." Now, it was really hard on my wife when we had the operation, because, just imagine this: here's our daughter going in to have her kidney removed. So, she comes out. Well, then I go in.

See, I'll just tell you: I've been so blessed in my life. And now, I'm at a time in my life where I recognize that all the things, all the experiences that I have had in my life—seemingly disparate, unconnected—have all been part of this mosaic that God has been working on, to get me to this point in history, where I am equipped and enabled to perform the mission that I've got, now, which is this Military Outreach U.S.A. At Military Outreach USA we work with houses of worship to help the veterans in their congregation. It has become a national effort.

When South Vietnam fell, I wasn't surprised. I was extremely disappointed in our government for not standing behind the South Vietnamese. They weren't asking us to send more troops; they were asking us to live up to the Paris Peace Accords and to continue to provide them with equipment. Then, it was over. We all knew that, all of us who were there. There were a lot of loyal South Vietnamese soldiers and civilians who had served and worked with us. They fought and died for their country, and we deserted them.

I saw what our government was doing, and I felt very bad, almost ashamed, that we did that to our South Vietnamese counterparts. I've always wondered what happened to my counterpart. The guy was a combat veteran on the staff at II Corps in Pleiku. We lost 58,000 soldiers, in support of them. They lost many more than we did, many times over. They weren't asking us to come back and fight; they were asking us to do what we said we would do. And we didn't.

For many years, I avoided going to the Wall. I don't know why, but I just avoided going. When I was promoted from colonel to general officer, I attended the general officers' orientation course. For one week, they invite the new general and his spouse to come to Washington, DC, for orientation. It was pretty high-level stuff. It was called the "charm school," by us. And, so, every day we were there, they had great events planned.

One evening's schedule was devoted to the Wall. I couldn't avoid it, I had to go. There were 40 or 50 of us. We went to the Wall, and it was beautiful. It was summertime. So, when I got there, I looked up the names of numerous guys that I knew. And that was very cathartic, I would say, for me, regarding just being able to bond back with the guys that I knew. I'd lost a friend who was with me at the University of Illinois. He was with the Americal, and he was killed. My roommate at Fort Lewis shipped over before I did. Remember that Life Magazine article, that one issue that had the pictures of all the guys who were killed in one week? He was one of those.

I only lost one guy in four months. He was my RTO. I had guys wounded, but I lost only one guy. He made it all the way back to Walter Reed Hospital. And then he died. I couldn't believe it. If you made it back here, 98% of the time you were going to make it. And he died, here.

I'm proud to have served in Vietnam. I believe that we should have been there, and I've never regretted being there. I've been so honored, too. People ask me what's the highlight of my career. The most important one was when I commanded my infantry company. That's the epitome for any infantry officer.

In Vietnam, we had generals and admirals that did not stand up and tell it like it was. And look at McNamara. Twenty-five years later, this guy writes his book, and he says, "Oh, by the way, I knew we weren't going to win the war. But I had to be loyal to Kennedy." What? How about being loyal to the men and women who were dying?

It was a special moment neither of us expected.

Specialist 5th Class Randy E. Richmond
402nd Transportation Co., 1st Logistical Command

I took a taxi from our small airport to the John Deere plant where my dad worked the graveyard shift as a security officer. It was just before midnight and I knew he'd be walking from the parking lot to the clock house at any time. I was home about three weeks earlier than my family expected me so I thought I'd surprise everyone. No one knew I was coming.

When Dad saw me get out of the cab, it was a special moment neither of us expected. I could physically see the gladness, relief, and love in him as we snatched a quick man hug there in the street. Then he took me into the clock house and introduced me to a couple of his co-workers. A little over 20 years later, my son saw those same emotions in my eyes when he returned from serving in Desert Storm.

1st Lieutenant Tucker Smallwood
MACV, MAT-36, Advisory Team 44

I went to three different colleges. I received an engineering scholarship to Antioch University in Yellow Springs, Ohio. Antioch is a liberal arts school; very expensive, very small. My test scores have always been impressive but I was not very mature or very disciplined, and now away from home I discovered girls, poker, beer, and guitars, and did not apply myself to my studies. After a year I was not doing well academically. I had the opportunity to go to college in Munich; my father was in Germany at the time and thought it would be wise to keep a closer eye on me, so I went there for the next two years. When I finished those two years in 1964, I came back to the University of Maryland to complete my education. I wandered into a TV studio one day and had ... what would you call it? . . . a moment. I just realized this is what I am meant to do and this is what I should be doing. I had studied electrical engineering at Antioch, and international relations in Germany. I graduated with a degree in television production and a minor in German, and I went on to direct TV for NBC in Baltimore. I was their first director of color in their history.

In June 1967, after working at NBC for six months, I received my notice of induction. I had injured my knee playing football for the University of Maryland and had been classified as 1Y (Registrant available for military service, but qualified for military only in the event of war or national emergency); however, the casualties in Vietnam were increasing and the doctors at the induction physical said, "Your knee clicks every time you flex it." I said, "Yes." "But you can walk, right?" and I said, "Well, of course I can." "Well, you let us know if you have any trouble." And I was on a bus that night for Fort Bragg.

My time in the Army was not particularly typical. Within 24 hours it was very clear to me that I was either going to have to get some rank or I was going to spend my next two years in the stockade. I never had any patience with being messed with and I've never had any patience with bullshit, and I was subjected to it very early on and I realized intuitively, shit rolls downhill. I went to basic training at Fort Bragg and infantry AIT at Fort Dix. I was a platoon guide in both basic and AIT. I was busted in the eighth week in basic for playing poker and I was busted in the fifth week of AIT for telling an NCO to get off the road while we were on a tactical move or else I would butt stroke him with my M14. And so I was busted.

My father had influential friends within the State Department and it was "arranged" for me to be assigned to a documentary production the Army had called "The Big Picture." With my background, I was a perfect fit; however, I had fallen in love with soldiering. I was a city kid who fell in love with the physical demands of navigating, enduring, persevering, and all of the things that had not been a part of my childhood. It was just a coincidence that we had a candidate whose father was a general, and he spoke to his father about me. Strings were pulled, which is the way things work in our world, and I found myself with an opportunity to get a commission, which meant I would have a little rank, I would have a little power, I would have a little more money, and I would have a little bit more choice.

And so, when the opportunity was created for me to go to OCS, I took it. My father was not pleased with me but I had my own mind made up. In six months of OCS I was on restriction for 22 out of 26 weeks. I had two Class I and four Class II infractions: generally, half of that is enough to get you kicked out. But I was also on the Commandant's List when we graduated. I was not a barracks soldier but I was very, very good in the woods. And so, when I graduated, my orders were initially for Special Forces or the Old Guard in Arlington, but they pulled those orders to make me a TAC officer, because at that time that was the highest priority in the Army: to create more infantry lieutenants. For the next six months I got to relive the physical rigors of OCS and train young infantry lieutenants. It was not fun, I was not a nice person, but I took my work very seriously and at the end the Army asked me, "What do you want?" and I said, "I want jump school, ranger school, special warfare school, Vietnamese language school, and I will not work with Americans." They said, "No problem." There wasn't time for me to get to ranger school but I got everything else I wanted.

I arrived in Vietnam in June 1969. I was supposed to be an advisor to the ARVN Airborne but a "ring knocker," a West Pointer, had talked someone into taking my slot. He had pull. "What else you got?" I asked. Well, our new policy of Vietnamization had created something called a MAT. That was a five-man Mobile Advisory Team, and I said, "Yeah, that sounds good to me." And I have never regretted that decision for an instant.

I was assigned to MAT 36, Advisory Team 44 in Gia Dinh Province. We were advisors to the local regional defense force, essentially citizen soldiers. We were a five-man American team which was supposed to be commanded by a major with a captain as an XO, but in Vietnam the reality was the CO and XO were both lieutenants. People were just getting wounded too fast. The man I replaced, my CO, was in-country for an entirety of four weeks. He was there two weeks before I got there and then he was there two weeks after, when he took an AK round to the buttock, and I took over command. The five COs before me had less than a total of eight months between them all. It was a very dangerous place. I did manage to set a new record for individual command longevity at MAT 36—22 weeks.

Coming Home

On my last mission I was seriously wounded in the neck by a piece of shrapnel. I was dusted off to Saigon, and sometime after arriving at the hospital I was pronounced dead, but one doctor would not give up on me and he brought me back and kept me alive. I spent about three weeks there, and when I was strong enough I was flown to Camp Zama, Japan. I spent two weeks there and was finally strong enough to make the flight to Walter Reed, where I arrived in mid-October 1969.

The reality was that the doctors didn't know exactly how long I had to live; I shouldn't be here. The work that they had done was miraculous. but there was simply no track record with my injuries. because people with my injuries don't live, so there was no history. About two weeks ago I had a note on my blog from the surgeon at Walter Reed, Dr. Rich, who had performed some rather radical surgery on me, grafting a vein from my thigh into my throat and replacing the work that had been done in Vietnam because it had since clotted. He told me that after all these years he had remembered me and he wanted me to know he was happy that I had enjoyed a good life, and my records went into the Vietnam vascular registry. They are part of history now. For years after that, in the early '70s, I would return to the Walter Reed every so often so that they could do a 5,000-mile check-up on me. They would walk me around the wards and approach other guests: "Let me tell you what happened to this lieutenant. What do you think? What do you think his odds were? Should he be here walking and talking?"

While a patient in Saigon and Zama, I spent a lot of time in reflection. As I lay in that hospital thinking, "Well what is next?" I realized I knew technical things. I had been a director but I didn't know how to talk to actors. I thought, "Well, if you study the process, perhaps you would become a better director." And I loved the work I'd seen on the stage, so I contacted Sanford Meisner at Neighborhood Playhouse. He was part of the great trilogy; Meisner, Stella Adler, and Lee Strasburg. He said, "I want you here in September," and I said, "I will be there."

I was a patient at Walter Reed until the spring of 1970, but I was essentially healing so I could survive the subsequent surgeries that needed to be done. I was allowed a week or so of convalescent leave to visit my family in Germany, and when I was finally released from Walter Reed I reported to my next assignment, which was at Fort Belvoir. The engineer OCS is located there and engineers are tasked to fight as infantry when needed, so I was to become an instructor of tactics, patrolling, ambush, how to function in the jungle, how to perform air mobile assaults, and whatnot. I enjoyed that work immensely and did that through the summer.

In September I resigned my commission. I never really thought of making a career out of the Army. Not really. One, I had been in their reality of it; even before I arrived in Vietnam I had spoken to enough officers to know OUR WAR was fucked. It was not what it was cut out to be. We should not be there, but I was willing to help the people achieve their own self-determination. I did not want to be with an American unit because I was a young black officer. I was a hard-charging infantry commander and I did not want anybody rolling a grenade under my bunk or putting a round in the back of my head because I wanted to get the job done. I wanted to be with people who wanted to fight the war, who had a vested interest. And, as I said, I have never regretted that decision for an instant.

After I left the Army I drove to New York and began to study acting. I worked as a waiter and I studied. And I fell in love in the process. It was very beneficial for me because I had been shut down spiritually and emotionally for some time. My feelings, my passions, were not useful as a commander and in the process of studying acting I learned to re-access what is inside of me, how I feel about things. And that really was the humanizing that saved me, that gave me back my real life, and I have loved my work for all these years.

Home

I did very well for ten years. I was very, very successful, and then elements in my life began to unravel and I followed the path into depression that many of us do. That's why they call it posttraumatic stress disorder. Some of us do well for a day or a week or a month or a year. Some of us break down after 10 years, some of us break down after six months; it just depends upon the circumstances of your life. Something triggers something and then it begins to dissolve. For eight years I wandered through my life, filled with depression, very low self-esteem, and self-destructiveness, essentially because I felt guilty having the life that I had when so many people I cared for were in the ground.

I have never been a joiner but I went to the dedication of the Wall in 1984, and I came back to New York and they declared a parade, a welcome-home parade for Vietnam veterans, and I chose to participate. I am not sure why; I just felt it was something I wanted to be a part of, and I met men that day that became my friends for life. And from that day I became involved in advocacy. I am still troubled, and two of those guys that I met—brothers—one day handed me a book. They risked our friendship because they realized I might not react well, but I read the book and I recognized myself in these interviews. The book is "Long Time Passing" by Myra MacPherson. I have since connected with her. I recognized myself and I said, "My gosh, that's me, that's how I feel, that's how I behave." And so I sought help, and Dr. Victor Defazio took me in. He had been to Vietnam before he received all his training in psychology. He was working with Russian psychiatrists to develop therapies for the Russian soldiers in Afghanistan who came home with the same problems: drug addiction, guilt, self-destructiveness, unhappiness, PTSD. And over a period of months he helped me regain my self-esteem and my self-worth. Part of this therapy was to write down everything you could remember—don't judge it, just write it down. And I did, and that became my task and out of that comes my writing. I discovered my voice as an essayist. I like good fiction. I don't write it well but I can describe events that happened to me and look through my own lens and say, "Yes, that is true, that is what happened, that is so."

I am a resource because I have a passion for advocacy. I am a public figure and have been for more than 45 years. I am recognized all over the world. Since I was no longer a private citizen I was less reluctant to say to the world, "I live with PTSD. You know me as an actor or an alien or this boss or this commander. I am also a Vietnam veteran that lives with PTSD." There is much stigma attached to PTSD. I wanted to put a human face on the condition. And because of that I have been in contact with countless people through telephones, emails, and personally, giving them a space to share their concerns, to offer guidance when I think that can be constructive. It is a fundamental part of my life. I made that pledge when I was saved by Dr. Victor Defazio. I got lucky. There are three obstacles. The first one is that you have to realize that you are in trouble, because we can live in denial for a long time and we are trained to drive on, to function. Second, you have to be willing to accept help, and, third, you have to find it. So those are big hoops to jump through. And I made it and I said I will do what I can to help any veteran, no matter the war, to get the help that he or she needs so that they can recover the quality of their life.

I don't know whether my spirituality changed because of Vietnam or I just became more informed. My father was a Lutheran minister. I was born in the Christian church; I was an acolyte. I went to college, I read Freud, I read Allport, and I perceived myself to be an agnostic. My dog tags said, "No preference." I am many things, but I have never been a hypocrite. Most people would have declared being a Catholic, or Jewish, or whatever; my dog tag said, "No preference." I have my dog tags, I am looking at them right now. They are in a little case about my desk with some medals and some photos. I am spiritual; I am not religious. Over time I have studied Buddhism. I have meditated for years. I am spiritual. It was a slowly evolving process. My belief systems were a la carte: I took what made sense to me, what resonated with me over the years after meeting many different people.

I have been to the Wall many times. Privately, generally late at night, but I did participate in the 25th anniversary and the reading of the names. That was very moving, I was grateful to be a part of that. For three days and nights we read the names. That's all, just read the names.

The Wall is unique in its simplicity and it's a sacred place for us, it is a place where we give ourselves permission to feel and grieve. A lot of people are visiting the Wall now, and there are a lot of us that just can't do it. The Vietnam experience was both personal and universal.

I haven't experienced any health issues from Agent Orange so far. I was in the water constantly: I was wet more days than I was dry. It was certainly all over my area of operation. I drink, I smoke, I have challenged my body in my years, but, knock on wood, what I haven't had to deal with is any issues from Agent Orange. I count my blessings. So far so good, that's all I can say. Dioxin was all over the place, it was all over the area I was in, it was in the water, and I was in the water constantly. I've dodged that bullet so far.

It was just a question of time when South Vietnam would fall. I didn't take it personally. The reality is that Vietnam was a political war, a proxy war, and the people of North Vietnam believed in their cause and the people of South Vietnam did not.

Looking back, I think Vietnam was a life-altering experience. It is what happened, it is what was in the space or my destiny or however you want to look at that. I am entirely who I am because of it. I thought about my hierarchy, if you will, when I was writing something for my theater and I realized:

A) I am a veteran,

B) I am an artist,

C) I am a man of color.

I am a man as opposed to a woman and I am a man of color. Those three describe my personal hierarchy.

The war was what it was. There are lies, damn lies, and statistics. I know what I took away from it. I have no interest in arguing with anybody else about it. I am not going to persuade you, you're not going to persuade me. I know what I know.

My trip back to Vietnam was short perhaps in human terms; it was long enough for me, it was just what I needed. I am so grateful that I chose to do it. I am healthier and happier and more stable because of that visit. It just reaffirmed my affection for the people, for the culture, for the country. Would I go back again? I certainly want to go back. If it is meant to be. I just hope that perhaps my life and my work will take me there. I am certainly willing and will look forward to it.

Specialist 4th Class Dennis Haney
1/6th Infantry, 198th Light Infantry Brigade, Americal Division, U.S. Army

Back in the olden days, when folks were proud of where they lived and the history related to the surrounding area, there were places like Fort Wayne in Detroit that most kids either visited on a class trip or a family outing. We all marveled at the stuff in the fort and fantasized about playing there, even though there was no way we could do so. After all, we were raised watching John Wayne and all that unbridled patriotism in his movies, not to mention all the countless hours spent in a toy arms race with the kid down the street, as we waged our toy battles in each other's backyards. We idolized our fathers who were all in the service during the "Big One," and the end result was we felt that we were well versed in the art of warfare.

In the spring of '68 I got my notice to take my physical for the draft and saw the handwriting on the wall. The idea of being in the military for two years wasn't very appealing, but, on the other hand, my father and uncle served, my older brother went when called, and even Elvis went when it was his time, so it seemed normal to become part of the long olive drab line.

Although the anti-war protests were intensifying, I figured that my die was cast, so there was no use trying to get around it. Finally, I got my greetings, and while I was on my way to Fort Wayne, I did some reminiscing about all the good times spent while visiting there. Right after I walked through the door and took the oath, all those pleasant memories went south very quickly, so much so that I haven't been near the place since (LOL!). We got a lot of practice with "hurry up and wait," and by the time we were herded into the Continental Trailways bus to take us to Fort Knox, my "US" number was forever etched into my head. As we got closer to our destination, I saw that the dirt was red instead of the gray clay I was used to. Little did I know at that point that in the next year or so I would see more than my share of red dirt.

Basic training was just that: "We'll make a soldier out of you one way or another!" We learned helpful stuff that would come in handy later in life, like how to make a bed, polish boots, or how to crawl around in muddy sawdust. The only thing worth mentioning other than training with M-14's is a story that goes back to grammar school. There was a kid who enjoyed hassling me a lot back then and, lo and behold, he was in my platoon. He pretty much picked up where he left off whenever he saw me. The DI saw that, too. I didn't want to do anything that would make my hitch go beyond two years, so I took the hassling in my stride. One day we were training with pugil sticks that were padded on both ends, and the DI matched me and that guy. I saw who I was paired with and figured that I was going to get my clock cleaned pretty good. Sure enough, it started out that way. He gleefully delivered blow after blow. Things were going as predicted until he finally hit me hard enough to hurt me. At that point, I got PO'd and went after him like the kid in Christmas Story. I had blown a fuse and ended up being dragged off this guy who had tormented me all those years in school. After his whooping he avoided me, and I never saw him again after basic. Looking back after all these years, I now realize that the DIs knew that the quickest way to learn and remember something is pain, and since their time with us was brief all they were trying to do was improve our chances of survival in a very hostile environment. If I ran into one of those guys today, I'd buy him a drink and thank him for doing what he could to help save my life.

Basic went by fast enough, and towards the end we got our orders for advanced training. I, like most others, was blessed with a trip to Tiger Land at Fort Polk for advanced infantry training. The training cycle was shortened for us to go home for Christmas and be in-country just before Tet, in case they pulled an instant replay of 1968. The disappointment of not getting what we put on our dream sheets quickly went to the wayside, because our fates were sealed as grunts.

Coming home for the holidays was bittersweet. We were all marked men in our Class A uniforms and our short hair. Our false bravado was just an attempt to try and mask how scared we really were. I sincerely felt that I wouldn't survive my tour, but I gave no hint of that to my loved ones. When my leave was up, it was time to go off to war.

I reported to Fort Lewis, and there was something surreal about walking around in the snow with jungle fatigues on. Finally, I was on a plane heading to the Orient. The flight seemed to take forever, but we finally got to Cam Ranh Bay. I went from snow to an almost unbearable combination of heat, humidity, burning, whatever, and decomposing jungle. What a terrible aroma! I had the good fortune of being chosen for shit burning detail. It was extra special because of the numerous dietary changes and queasy stomachs stemming from wondering what

was going to happen next. All this made the barrels a real joy to handle. I only had that detail once, but I felt that I came into country on the right foot.

Somewhere along the line, I was offered both NCO and Warrant Officers school, but I would have had to extend my hitch another year, which was totally out of the question. I was assigned to the 1/6th Infantry, 198th Light Infantry Brigade, Americal Division out of Chu Lai, but I worked out of LZ Bayonet. When I got to the 1/6th, they asked for volunteers for Recon. No extra hitch time was involved, so I raised my hand even though I didn't even know what Recon was, but I was about to find out directly.

The first night at LZ Bayonet I and two other FNG's were issued rifles and put in a fighting position right next to the HQ hooch. We had no idea that we were deep inside the compound, all we knew was that we were new in-country and scared to death. There wasn't any problem staying awake that first night, especially after the bunker line did some H&I fire that no one warned us about. We thought the base was under attack, and we waited for the gooks to come and get us. Although terrified, we didn't leave our position even though we expected the worst. The next day we saw how far we were from the wire and the word "embarrassed" didn't come close to describing how we felt.

That first ride on a helicopter set the stage for many others—the noise, vibration, and the wind blowing around the cabin rivaled any ride that an amusement park had to offer. I didn't realize we were coming in hot (after all, I hadn't done anything worth getting shot at, lol!), and I was waiting for the ship to land when someone very unceremoniously pushed me out the door. Once that was over with we began to hump at a good pace and were nailed three more times that day (I was cleaning my rifle the third time and I think I broke the land speed record for putting it back together.) I hoped that every day wasn't going to be like the first one was, and they weren't. I've always felt that I was better off seeing so much action right off the bat because it took all the guesswork out of the rest of my tour. One thing for certain, though; being in combat wasn't anything like a John Wayne movie.

Thanksgiving came and went. Christmas rolled around. I missed being home for the holidays but I only had a few weeks left on my tour. I would be going home soon and that was the most important thing to me. New Year's at midnight there was a fire power demonstration that outdid anything I ever saw before or since. Two weeks short! On my birthday, I became a Single Digit Midget (only 9 days left to go). That's all that mattered to me: getting back to the world.

When the day came to leave it was same-same, like when Radar left MASH. All my friends were either on duty or in the bush. I did take one last look around, knowing that I would never pass this way again. A flood of conflicting emotions passed through my head. In spite of a tremendous relief knowing that I managed to survive my tour, I also, equally, felt intense regret that I was leaving my friends behind, but they too had to serve out their tour like I did. It was a year I'll never forget, that's for sure.

Coming Home

The trip home is hard to describe. It was a gamut of emotions like relief, disbelief, shock, and the anticipation of finally going home. Although we were a bunch of guys who had just shared a common experience, we all knew that we were in transit so no real bonding took place. Mostly we were just happy that we weren't getting shot at and there were no booby traps around.

The only thing I recall about the trip home was flying over the Arizona Memorial as we landed to refuel in Hawaii. It was a sobering sight even at the altitude that we were at, especially now that we were vets who had seen war.

Once at Fort Lewis we were warned about the protesters. "Protesters? We didn't do anything to them, why are they hassling us?" And they warmly greeted us with burning bags of dog crap. They thought that they were so self-righteous, while we were considered scum of the earth. One questioned that train of thought considering that many of us were fresh out of the jungle where life was cheap. If we had gone after them to retaliate they would have been dispatched and we then would have gotten our steak dinner as though nothing had happened. They were fortunate that we rose above their ignorance.

I wore my greens home because I had the stare and I knew that I still carried my Detroit attitude. I was avoided at Sea-Tac, which was okay with me, and once I was at the Detroit airport there were no problems, mainly due to an old Detroit adage; "You don't start no shit and there won't be no shit!" I called home to let my folks know that I'm nearly home and decided to take a cab rather than wait over an hour for my family to get there. Getting into some civvies was paramount in my mind (after my family of course). Since it was the house I grew up in it wasn't much of a shock arriving there. I had left and come back a number of times, so the act of doing so was somewhat routine.

This next part is something that I haven't thought about in a long time. Nearly two years had passed since I was drafted and everyone had gone on with their lives while I was in a time warp for a year. They were eager to get me current on what had transpired in my absence and all the stuff that happened to them. Except for the girl next door, none of the others could care less about what I had been up to. They did their best to keep all our conversations about them. I thought that surviving a life-or-death situation might be worth listening to, but such was not the case. Their lack of compassion put a distance between us that is still there today. As I watched the Army-Navy game I got sort of a revelation on that. We were all at an age that we all constantly had to prove our "adultness," no matter how many chicks you laid or how many bars you closed. Going to war and coming back alive is, hands down, the truest proof of manhood that there is. Believing in the war or not had nothing to do with it. They were not only out-shined but had their fragile egos roughed up a little too.

It was a shock coming home to find that the concept of "The World" that we all clung to and kept us going no longer existed. It had moved on with the rest of the world in our absence. Getting such a warm reception from the protesters didn't help any, either. It's no small wonder why so many of us chose to grow our hair long and try to forget that place and all that happened there.

Home

I wanted to talk to another grunt in the worst way, but the only one I knew had moved away. I felt very alone not being able to talk to someone who understood. It was obvious to me that I was screwed up, so I went to the VA. They were "peaches & cream" with me until they got their hands on my medical records, which they promptly lost, and told me that there was nothing wrong with me. I tried to get a copy of them from St. Louis but was told that they were conveniently burned in a fire. Feeling very betrayed, I said, "Fuck the system!" and that was that.

It didn't take me long to realize that it wasn't a good idea to put my service on a job application. I didn't mention it to anybody (thank goodness my hair grew out.) Being a draftee, I figured that it wasn't my idea to go to 'Nam, but the citizens had different ideas on the subject. People in general were real shitheads when it came to us vets. The media portrayed us as a bunch of fiends and were very successful at doing so.

I couldn't even talk on the CB radio without some bozo bitching about me using military jargon. I was our command RTO, and that jargon saved our bacon more than once, and I couldn't see me not using it because some oaf didn't like it. I tried to join the VFW, but those old farts didn't want anything to do with us losers. Vets being down on other vets came out of left field, which didn't help our sense of alienation much. Finding out years later that due to air mobility we saw more contact in just a year than most WWII vets during their whole war really diminished my respect for them. As time wore on I grew more and more solitary and had less and less to do with the general population. Listening to them whine about their chicken-shit problems convinced me that they had no clue of what real adversity is, like being in heavy contact. The years passed by and I did every dirty job that came my way (while in a firefight I swore that I didn't care what I did for a living as long as I wasn't getting shot at). The only problem was that I couldn't keep a job for any length of time.

About 15 years ago it got to the point that life in the jungle seemed to make more sense than civilization, so I went to the VA and got into the system. It took them two months to convince me that I had PTSD, because I always felt that I could beat the demons (it was a blow realizing that I couldn't), but now was the time to man up and accept my fate. The more I met with my care provider, the more I regretted not knowing about it sooner. Oh, well! Xin Loi! At least now I know what I'm up against. Thus far I have no issues with Agent Orange.

When Saigon fell, my initial reaction was great sadness that all those brothers fell for nothing so that somebody could make a lot of money from our sweat and blood. We didn't fight to help big business get richer; we fought to help each other get home. There was some regret for the locals, but, on the other hand, they didn't work too hard at taking care of their own dirty work, so they got what they deserved.

I wasn't religious when I got there; I figured that if I were to find religion that would be the place to do so. No matter how tough things got in the paddies I never cried out to God for any intervention, so I guess that there are a few atheists in the foxholes, after all, in spite of popular belief.

I've been to the portable Wall and what bothered me the most was not remembering the names of those I knew who had fallen.

Just lately I came across a picture taken on leave just before I went over. It's of me and two of my friends gleefully flipping the bird to the camera (we thought we were so grown up back then, lol!.) I went to war an immature, naive kid and came back a man old beyond his years. Learning that you're expendable at twenty is a tough pill to swallow.

In retrospect, I'm amazed at the feats I performed, like crossing impossible terrain (although I still cringe at the thought of walking on a rope bridge) with a very heavy ruck on my back. Never in my wildest dreams did I ever think that I could do such things, but yet I did. I saw countless instances of individual heroism because a buddy was in peril, and I hope that all my Recon brothers have had life smile upon them, because they certainly deserve that.

It's too bad that sometimes one has to grow old to really appreciate the words, "Duty, Honor, Tradition." In my youth, I would have never believed that, but now, as an old man, I am proud to say that I was a part of that long olive drab line.

I had this recurring nightmare. I was needed and went back, and I got killed.

Specialist 4th Class Gary Sieroslawski

A Battery, 7th Artillery, 8th Battalion, Second Field Force Artillery

I am from the northwest side of Chicago. I was drafted into the Army in August 1968. I was changing schools and going part time; I ran out of deferments. I got the letter from Lyndon Johnson saying, "You're invited to join." Actually, it did say, "Greetings."

My brother was going to school, couldn't find a job, and had financial problems, and, finally, he said, "Screw it," and joined the Army. He went to radio school and ended up being stationed at the foot of the Alps. Good place to be.

My basic training was at Fort Polk, Louisiana. I had no idea what job I would have after basic. At the end of the eight weeks they called everybody together and gave them their orders for their next duty station and school for their MOS. They never called my name. They didn't have my orders. It looked like they lost them. I had to go running from building to building to find my orders. I ended up going to Fort Sill in Oklahoma. My MOS was to be 13A10: Basic Field Artillery.

So they bussed us directly from Fort Polk to Fort Sill in November 1968. We did not receive any leave time between basic training and Fort Sill. The artillery course is normally an eight- or ten-week course, but we finished it in seven-and-a-half weeks so everyone could go home for Christmas.

We received our orders the day before we left to go home on leave: Vietnam. I wasn't that informed about Vietnam at the time. I had seen it on TV and I had a faint inkling that I'd probably be going there. I sometimes thought I wanted to go there, just to see what it's like because of all that's been said and written about it. But I was happy I was in the artillery and not the infantry.

I went home on leave for ten days just before Christmas. I didn't tell my folks beforehand that I was going to Vietnam. I was going to tell them when I got home. In fact, I think I told them in the car on the way home from the airport because they wanted to know.

My girlfriend's father was a belly gunner on B-17s in World War II. He was shot down and parachuted twice during the war. When he found out I was going to 'Nam, he said, "I got one word of advice to you." I said, "What?" He says, "Keep your ass down because you make a big target."

I flew to Vietnam out of Oakland on a charter flight, like most did. It was Transworld Airlines or Trans American or something like that. We flew to Hawaii, then Wake Island. I got to see Wake Island, and that was awesome. There was a hill, and a couple of other guys and I climbed it to see what it looked like. The runway literally went from one end of the island to the other. I couldn't believe how small the island was. We then flew into Bien Hoa.

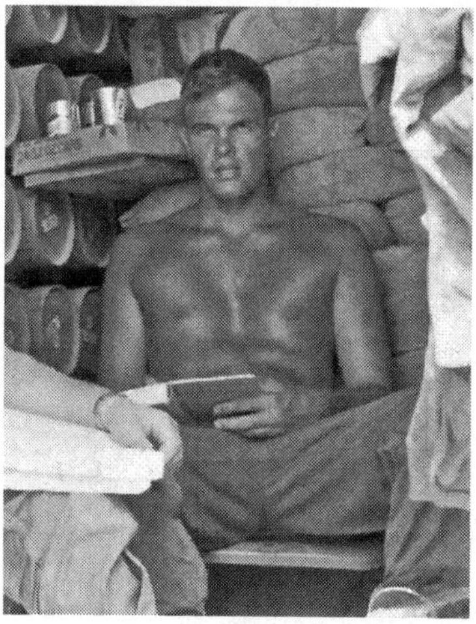

My first impression of Vietnam was that it was hot and stinky. The first thing that hits you when you get off the plane is the heat, and the next thing is the smell. You can't describe that smell. People complain here and think it's hot, and I say, "You don't know what hot is. You have no idea what hot is."

We boarded a bus to Long Binh for processing and in-country training. The bus driver gets on, and he has an M-16. We weren't armed. The driver turns on his radio as we're driving to Long Binh. He is listening to AFVN (Armed Forces Vietnam Radio). It was just strange. You can see the villagers looking at you as you're coming through at night, driving through town. You can't get out. The windows are all screened in, and it was like, "Oh, my God." It was an unsettling feeling.

I was assigned to A Battery, 7th Artillery, 8th Battalion, Second Field Force. We had 175s and 8" guns. We rotated between Bien Hoa, Xuan Loc, and Bien Loc. Bien Loc was the worst of all three because that's the place that got hit all the time. The first time I went in the mess hall at Bien Loc I looked at the table, and it looked like Swiss cheese, because of all the shrapnel holes in it. They used to serve dinner at four o'clock because by six you started getting incoming.

We moved around a lot. Various fire bases. I once saw an article that said, "A fire support base is like the Old West where you have a fort, and it's out in the middle of nowhere. The only difference is, instead of having bows and arrows and rifles, they got AK-47s and RPGs." That is how I spent my tour.

The Army had an Early Out program. At the time, things were relatively calm and peaceful, and other than the long hours it really wasn't so bad. So I said, "Well, I'll take the Early Out." I had to extend my tour for two months in order to get it. I did that mainly because I saw what they did to these guys with five or six months left after they came back. They gave them all the shit jobs. I said, "No, I don't want to do that after being here."

And then about a month after I had extended, things hit the proverbial fan. It was a couple of days here, a couple of days there. We were on the road a lot. We even worked with the Aussies up in Nui Dat. They're another story by themselves. I've never seen anybody drink as those guys could.

Coming Home

I was at Bien Luc when my orders came through. I had to go to headquarters in Bien Hoa to clear post. Then I went to a holding company and waited for the plane that would take me home.

You sat there and waited. Every time a flight would come in, they'd go through the roster of who's going on the next flight, and they would call off the names of the passengers. You basically didn't sleep because you didn't want to miss your name being called.

We flew to Okinawa and then Anchorage. This was about April 2. We had to get out of the plane and go down the ramp in our jungle fatigues. We had spent the last 14 months in 100+ degree weather, and now we're hitting -30 degree weather. It was cold. Some guy said, "If you're over 18, the bar is that way." Well, everybody was over 18. Guys were buying bottles to bring back on the plane. That was a happy flight from Anchorage to Oakland.

Buses were waiting for us when we arrived in Oakland. It was nighttime, but protesters were out there with their signs and throwing stuff at us. Welcome home. That's when we knew things were different.

When I went to Vietnam we landed in Hawaii, and there was something wrong with the plane, so we had a delay for a couple of hours. We were in jungle fatigues walking around the airport. Nobody would talk to us. They all knew where we were going. We got a cold reception going over there and a hostile reception coming home.

When we were in Oakland Army Depot we were given a free steak dinner. I mean, there was a guy that said, "Wow, free steak dinner." I told this guy, "What are you talking about? How many dinners have you paid for in the last year?"

It took a couple of days of processing because I was getting discharged. And in early 1970 the air controllers were having a work slowdown, so there weren't as many flights taking off right then. I finally got my papers, discharge, and my muster-out pay of $600, $700 in cash; big money at that time. So I went to the airport, took a cab with a couple of guys. This was Saturday night around five.

They gave us our dress uniforms, the dress greens with slick sleeves, nothing on it. They didn't even sew your rank on it, because they didn't have any. There were so many guys going through, they didn't even bother. I had my ribbons and a couple of pins I was able to put on.

I went to the airport and told the attendant I was on Military Standby. "The first flight I can give you is Tuesday morning." I said, "No, no." This is Saturday night. "Do you have anything available? I don't care what the price is. Hell, I got all this money. I'll pay what the price is." She says, "Well, I have a first-class ticket left on a 12:30 a.m. flight." I said, "All right. I'll take it. I'll take it." It was 6:00 p.m. so I sat down to wait for the flight, and people would get up and walk away. They wouldn't sit next to me.

So I went to this bar at the airport and stayed there until 12:00 a.m., and literally their bartender was just going in a circle filling drinks. And all of a sudden,, it was about like 10:00 p.m. or 10:30 p.m., these four girls come walking in to go to the bar. They took one look, turned around and walked back out. The guys are, "Yeah, come on in." One of the girls gave the guy the finger and kept going.

I got in the plane and next to me is a gabby old lady. She says, "Where were you?" I was just sitting there with a suntan and a uniform. It was probably obvious. I was talking to her, and the hostess comes up. "Let me see your ticket." She wanted to see if I was flying standby, and I showed her my ticket. She says, "Fine." Ten minutes later, the same one comes by. "Let me see your ticket." She must have done this four or five times. I don't know if she's just doing it to bug me or whatever, but frankly, I was trying to sleep, so I just laid the ticket out on my chest.

I finally landed at O'Hare and my parents were there with my brother and his kids and my girlfriend (my wife, now). My mother says, "He looks really tired." My brother says, "Tired? Hell, he's drunk." Which is half true, I was half drunk and half tired! That was April 4, 1970, at six o'clock in the morning.

My mother wanted me to wear my uniform to church. But after the grief at the airport and the flight home, I swore I would never wear it again, and didn't. I did end up throwing it out after my basement flooded and it got ruined along with my pictures, medal citation, and other Army stuff.

Home

After I was discharged in April 1970, I really wasn't doing anything for a while other than drinking and stuff. Most of the people I knew moved on, married, were in college, or graduated, so I lost track of a lot of them. One friend got drafted around the same time as I did. He was in the Army and stationed in Alaska. We spent a lot of time together when he came out.

My wife graduated college while I was gone. I remember going to her sorority dance after I came home. She was an alumna and some of the people knew who I was because she told them where I had been. The first thing they did was ask, "Did you see any action?" I said, "Well, yeah. Hell, everybody sees action." And then that was it. They wouldn't talk to me.

There was nobody to talk to. Another vet was the only guy you could talk to. So I went back to school. I was going to Loyola when I went in the Army and went back there for the summer term. I was sitting around and didn't have a job at that time. I finished going at night under the G.I. Bill.

I took a speech class, and you had to give your history at the first class. You had to tell where you had been. I was sitting there with a suntan and short hair, which was a dead giveaway. Some of the people in the class turned away, and even the teacher was a little cold. He didn't want anything to do with me.

It was strange the year I came out. My girlfriend (later my wife) and I probably had the most fights we've ever had in our lives during that year. Everything would set me off. I would take her home when I got mad, and her father would get mad because I'd peel away from the curb. Sometimes things would set me off quickly. I got married in January 1971. Once I was married, going to school, and working, things started to calm down for a while.

I had sleep problems for the longest time. I had this recurring nightmare—I was needed and went back, and I got killed. I would dream that, and I would wake up, and there'd be a gook over my head with an AK pointing at me. That lasted a couple of years. After a while, it just got further and further apart. It very rarely comes back now

One of the things that got me right after I got married, I was sitting with my father-in-law (the veteran), my parents, and my wife. We were talking. Something was going on in the news about the war, and my mother says "Well, Chester," (my father-in-law) "he was in a real war." And I said, "Excuse me? They weren't shooting rubber bullets at me."

Another thing that bothered me was that after you come home, especially the first year, every time you see the news and they start talking about places that you've been . . . and you know that the guys that you left were still there.

I have some bitterness toward the VFW. A friend and I got out about the same time. We used to play tennis and golf together before we went in the service. We went to a VFW Hall after playing tennis to get a beer. We walked in, and this guy says, "This is for veterans only." And then I said, "Well, what are you talking about? We are veterans." "You guys aren't old enough to be veterans." And I said, "What do you mean we're not old enough? We're veterans. Do you want to see my ID?" He said, "I don't care." The other guy says, "Drink your beer and leave. You guys are losers." And we thought, "Well, this is one VFW hall." I don't know if it was in Glenview or wherever it was, someplace. So next time my buddy says, "I found this other one." Basically, about the same thing happened there. We went to two or three different ones, and the same thing happened in all of them.

The American Legion was the same way. I remember distinctly right after I got out, about six months or so, maybe the following winter, the president of the American Legion came out and said, "These guys from Vietnam are nothing but crybabies and losers." And this was our support. I found out, talking to some of the other guys, that the Korean guys got treated about the same way, too.

My hearing is an issue. I have had arguments with the VA about my hearing. I put in a claim for tinnitus, and they denied the claim the first time because they said I wasn't exposed to loud noises. In the artillery, and I was not exposed to loud noises! I'm putting it in again and appealing. We will see what happens.

Looking back now, I guess if they asked me to do it again, I probably would. It's that type of thing. We can see what we think they did wrong, and I think we were fighting with one arm behind our back type of thing.

It's like everything, it's a big political thing. We never lost any major battle. People don't realize . . . people think we lost . . . we didn't. My way of thinking is, we didn't lose the war: we left. The ARVN lost the war. We were actually winning when we left.

SP4 Glenn Knight
B Troop, 3/17 Air Cavalry

My family moved to the Panama Canal Zone in 1962. We moved back after I finished high school in 1971. The adjustment was tough for me so I decided to join the Army. I was trained as a UH-1 Helicopter Repairer, 67N20. The recruiter said that the chances were good that I would be sent back to Panama. Ha!

My training qualified me as Huey crew chief, and that was the job I ultimately held in Vietnam. After basic at Fort Jackson we were put on a bus to Fort Rucker. We arrived late one evening. As we were seated during our in-processing, a sergeant asked us how many crew chiefs were present. Hands were raised, including mine. "Good!" he says, "Your next stop is Vietnam."

Well, he was right. I did receive my orders and, contrary to what my recruiter said, I was bound for Vietnam, not Panama. I expected it but was still pretty numb when I finally received confirmation I was actually going.

I was assigned to B Troop, 3/17 Air Cavalry at Phu Loi in Military Region III. We were hunter-killer teams, and worked predominantly along the Cambodian border. I spent my tour hunting the enemy. When my platoon sergeant in Vietnam saw me he would always say, "One day at a time, keep on smiling, Knight." Only I never have smiled very much, not then or now. There's much more to the story, but it just wears me out. It has been a long and winding road that has led nowhere. Peace out.

Coming Home

My family knew I was coming. It was great seeing the family again. I didn't have any friends here because we had only recently moved from Panama. I remember my mom made me a big country meal. I sat down and said, "Pass the fucking salt." Oops—took a while to clean up my language.

When I returned from Vietnam, my sister invited me and a friend I served with to a party in Greer, South Carolina. The people at the party were high school students. I hadn't been exposed to the hate of the veteran yet, but I would soon find out what the public thought about us. At some point at the party, my friend and I got separated from my sister. There must've been 30 to 40 people there. My friend and I were sporting our military haircuts, and that attracted some attention. After a few questions, the word was out that we had just come back from Vietnam. It was quiet for a few minutes. Then a group of about eight guys approached us. Their words: "So you motherfuckers have been in Vietnam?" I replied, "Yes, we both served our tours." Then a few choice words let us know we were not welcome. They started approaching us. It was either leave or get our asses kicked. We ran for the car, and they chased us. This was our first encounter realizing we were no longer in a friendly land.

Shortly after this incident, I went to the unemployment office, trying to get compensation until I could find a job. The state government official sitting behind the counter started giving me a hard time. I charged the desk and was restrained. Needless to say, I didn't get any compensation. More government red tape, more hurry up and wait, more, "Bend over, I have a surprise for you!"

I finally got a job at the Greenville-Spartanburg Airport fire department. It was mostly veterans working there, so I had no problems at work. I think the fire chief was a Navy veteran.

I started dating a girl, and we moved in together. The station captain on my shift and I were good friends. One day I came home and he was leaving my house. I went inside and asked my girlfriend what he was doing there. She told me he was looking for me. This was my good friend. I told her I did not believe her. He was after something else. I had a .32 snub nose pistol. I grabbed it, got in my car, and tried to find him. This is the first time I remember the loss of control at this level. Don't know where it came from. But it didn't leave.

Home

Everything between my family and I was distant, but that was okay, probably the best way to be. I had changed so much and didn't seem to fit in anywhere. I was a square peg in a round hole. The same is true today, always has been.

My girlfriend and I eventually became engaged and got married. Her parents refused to come to the wedding. That's okay, too. She worked at the airport at a rental car agency, and I worked for the fire department, with a part-time job on the side. We had a very small, two-bedroom house. Soon she was pregnant with my firstborn, who hates me now. In 1979 we had another child: it was my son. We lived in the Greenville-Spartanburg area until 1979, then moved to Dothan, Alabama. I was supposed to get a job with Hayes International, doing aircraft repair. After arriving in Dothan, I found out Hayes was on strike and they were not hiring anyone. On my way home I stopped by Michelin Tire and put in a job application. Three weeks later I was working at Michelin, building sport utility vehicle tires.

The work schedule was horrendous. I started working the midnight shift. I would work for seven nights, then I worked seven days on the day shift, then I worked seven days on the evening shift. At times, I thought I was losing my mind. Nothing in my life seems to have settled since my arrival back in the States in 1971. After trying to be a good father and failing miserably, mainly because of my temper, I began to drink. I would leave the house and say, "I'm going out!" I think the wife was glad to see me leave. I spent a lot of evenings in the bars, drinking.

As the kids got into their teens, they became adversarial, challenging me at every front, hell-bent on doing what they wanted. I was not having any of it. I did the best I could with a wife who didn't love me and kids that acted like they hated me. The inevitable divorce was on the horizon and soon to be.

After I was divorced, I spent much more time in the bottle, at home and in bars. Alcohol was my new escape. But it was only good as long as I stayed inebriated. Soon I met a girl that I worked with who also loved to party. It was a perfect fit. So we got married in 2000. Soon a Harley Davidson was in the picture, one for her and one for me. Bike rallies became a regular because there was a lot of drinking. But she grew up before I did. She stopped drinking, and I continued.

Soon, party guy became PTSD guy! Road rage, temper out of control, full throttle my only speed. Someone, please make my day. I wanted someone to help me release the pressure inside. Of course, my wife didn't know what to think. Where did this guy come from? I made an appointment with the VA, was evaluated and tested, locked up in the mental ward for a few days. Trust me, lay off the mental ward. It's just the Thorazine shuffle. They drug you, and that's it. All this, of course, because I told my psychiatrist of my plan for suicide. And I thought it out. "You have a plan?" "Yes." "You know where you were going to do it?" "Yes." I didn't even know who I was. What happened to the mild-mannered guy, pre-Vietnam? That guy is dead. I was very rough on my wife, my kids, my mother, my father, and my sister, anyone that I crossed paths with. In my neighborhood, a lady told me once that I was just a crazy Vietnam veteran. She was right.

Once on the highway during a road rage incident, a lady pulled up behind me and in the turning lane got out of her car. She came up to my car window. She asked me, "Are you stupid or crazy as hell?" I told her "I'm crazy as hell and if you don't step away from my car I will waste your ass right where we stand!" She left. Be careful who you screw with, they may be very crazy!

The night sweats, nightmares, and lack of sleep continued for years. Sometimes I would wake up and my bed would be soaking wet from nightmares. Sometimes I would wake up sitting up in bed swinging. My life turned upside down. My wife still stayed with me after seeing all this.

As everybody knows, the VA will give you all the drugs you want. The more drugs you are on, the less trouble you are. Once you find a good counselor in the VA, you can count on the fact that he's not going to be there long. Why? Because he's a good counselor! I've sat in group therapy and watched a counselor lead a conversation and go on about body bags, maggots, and exploding bodies on their way home from war. These quacks are just drawing a check. They don't give a damn about us. I am still taking anti-depressants and anti-anxiety drugs to help me sleep, and drugs to wake me up in the morning.

The guy who went to war is a KIA. He never returned. I asked my family about this a couple of months ago. I came home completely changed. Angry, seeking Vietnam-style revenge on anyone I thought had crossed my line! Who was this guy that looked like me but had changed from mellow to angry beyond words? The new me. How sad. My children became victims of an angry father and a dad who screwed their lives up. A father that spent nights in bars and finally divorced their mother. My firstborn hasn't spoken to me in months. Sorry, kids. They say this is normal for PTSD. That sucks. I can't get a do-over.

This is a very difficult subject to write about. It's about my failure as a man, a father, a husband, a son, and a brother.

I can only add one important thing to this: I have a wife that truly loves me. Prior to me she'd never been around the military and didn't know anything about it, but she understood that PTSD was not me. I asked her one time, "Why are you still here after all the crap I put you through?" Her response was, "Would you leave me if I had cancer?" I said, "Of course not." It has not been easy for her by no means, but she's still here and she loves me.

Side note; I go on Fort Rucker for mental health treatment, same place I was trained to be a crew chief. So I feel like I'm *still* serving! A lifer! Ha Ha. I didn't know when I signed up that it was for life. I thought three years and a wake-up. Will someone PLEASE WAKE ME UP? Enough already…

(Author's Note)

After this interview, Glenn spent 45 days in the VA Hospital in Biloxi, Mississippi. He now doesn't beat himself up because of PTSD but still has family issues to deal with. He will tell you that he has a much better understanding of PTSD and its triggers. He told me that he knows he will be forever changed by PTSD.

"Knowing your enemy is the only way to deal with it. I have always said I needed to be in a hospital. I just didn't know where. Everything in God's time. God bless you and all my fellow veterans."

Sergeant Kurt Robinson
1st Cavalry Division, B1/8th Cavalry

I was born in Detroit, Michigan, and when I was about a year or so old, my daddy took me away from there to Pottsville, Kentucky, where his family was. It was coal mining country. Anyway, when we arrived at his mother's house she said: "Well, you ain't bringing that little bastard in this house." I was a baby, and he took me to one of his sisters who kept me while he tried to get a job. But in the meantime he got stuck on this girl. She hated me. She was resentful: she wanted him all to herself, and she resented the fact that I came from another woman. They told me that my mother was dead, that she fell off a boat and drowned. Anyway, I went to Kentucky, and I bounced back and forth between his sisters and cousins for a while. Finally, when I was about five, he took me to his sister in Christiansburg, Virginia. She lived on a farm, and we got there early in the morning. I can remember my uncle had his feet up on this little potbelly stove, getting them warm. She walked in and said, "This is your home, boy." My aunt was a living terror: "If it weren't for me you wouldn't have a place to live!" So I became a farm boy, and I promise you, we never had any power equipment, everything was done with horses.

We killed five to six hogs at a time, and that would be meat for a month. We had cows, sold milk and chickens. I worked hard. She would yell: "If it weren't for me you wouldn't have a place to live. I'll put you in a reform school if you keep it up." There was no television. They had a radio on which she would listen to gospel preaching. My uncle's brother also lived in this little old farmhouse. His name was Oak, and my aunt would yell at me: "Stay away from him or he will cut you."

Saturday, you had to work. Sunday, we would have a little slack time after church. I was never allowed to leave the farm because she said if something happened to me they'd sue us and take everything they have. She was mean for no reason. She'd beat the hell out of me. And she had a son. I'm a kid, and he was ten years older than me. The son would just sit on his ass, and I'm the bastard. "If it weren't for me you wouldn't have a place to live." She was mean, cruel, hurtful, spiteful. One day he stole $17.80. I'll never forget it, she beat me in my face with her fists and told me how much she hated me and how I would rue the day that I was born. Well, I grabbed her fist, and I said, "Let me tell you something. You will never, ever put your hands on me again!" She screamed, "Get out of my house!"

I was 17, and I had had enough. I knew I was going to be drafted, and then Vietnam. There was no lottery then. I drove over to the draft board. My aunt was in the truck, and I told her I had to use the bathroom. I went upstairs, and I said, "What do I gotta' do if I wanna' go in? I gotta' go anyhow. What if I just go in and go now?" And they said, "You have to get a signature from your guardian." Well, I said, "You can look at her in the truck right there, she's sitting in it, she's got polio." Which she did, her feet were all crippled up, and she couldn't walk up these steps, so I took the paper and signed her name. "What the hell have I done now?"

I remember it was September, and it was hot, and she said: "Go to the mailbox, boy, get back, eat, it's dinner time." I went down there and there was a letter: United States of America, no less. "Greetings, you have been selected!" I said, "Oh, shit!" I went back up to the house and I told them I was drafted. Next thing I know I was in Fort Bragg, and then infantry training at Fort Jackson, North Carolina. I turned 18 in basic. Went to 'Nam as a grunt. I was assigned to the 1st Cav Div., B1/8 Cav 11B.

I was wounded on August 23, 1968, and sent to Japan where they patched me up and sent me back. I was a platoon sergeant. I was out in the bush in a firefight, and they pulled me out and said, "It's your turn to go home, get the fuck out of here. Get home. Run! Get on that chopper." We were down to 12 men. They all saluted me when I got on that chopper. I was so sad that I was leaving them for someone else to look after.

Coming Home

When I arrived in the States, there was a protest at Fort Lewis, Washington. They were name-calling, spitting at us, throwing stuff, and calling us names. We came through this gate in a chain-link fence and there was this girl, she looked right at me and spit right in my face. She looked me right in the eye and then she ran, giggling. She took off and ran and ran. I thought, "Man, what in the world is going on around here!" I was so pissed off. Two days earlier I was in a firefight. I hated it. I had no one to come home to. I had received a "Dear John" letter after about a month in-country. I was just sad. I didn't have a home to go back to. I drank all the way from Washington.

Changing planes, we were the subject of snide comments. People didn't want to be around us. I finally got to Christiansburg, Virginia, at three o'clock in the morning. I got off the bus in uniform. The old bus driver opens the luggage compartment and throws my duffle bag on the ground, gets back on the bus, says, "Take it easy," and off he goes. So I am left standing there. It's just quiet; you couldn't hear anything. Threw my duffle bag on my shoulder and I started out towards the farm. I walked and walked, and it took me nearly an hour-and-a-half to get there.

I got back to the farm at four thirty in the morning, knocked on the door, and the old woman looked like somebody walked over her grave. Duffle bag still beside the door, all she said was, "Well, you did make it." Then she immediately put five packages in my hands. She said it was something I sent home (my 1911). I slapped it together, and she says, "Lord have mercy, what are we going do with you? You might as well get in here and eat." She was cooking eggs and lard. I made myself eat, and she informed me if I wanted I could take my meals there but had to sleep out on her porch on a metal glider. Her son and brother told her I'd killed babies and would kill them in their sleep. Now, about this time the sun was coming up. I was standing on the porch, the morning sun hitting me in the face. And right then I realized I was never so alone in my life. I wanted to be back in 'Nam with my guys.

I stayed there about a month, something like that. My friends shunned me. They acted like I had the plague. They were scared to be around me. They heard so many war stories that we're baby killers and all this damn shit, with rape, pillage, and plunder. They said I changed. It wasn't the same. Home was a foreign country.

I had some money so I bought a car and a 1950 Panhead motorcycle. I joined a motorcycle club; the Gypsies (now the Pagans). They were mostly vets. I wandered all over for a long time. I rode it across the United States a couple of times. I had never in my life heard the words "I love you" come out of anyone's mouth. But, thank God, I made it.

Home

I went through a period in my life where all I had was me and the car, and I'd park it and leave it. I also had my motorcycle, and, you know, that's all I had. I went down to Louisiana and worked for a while, and then Colorado. I met my wife. I call her my cowgirl. She's a little Cherokee Indian. When we met, it was her birthday. There was a party, and she came walking by, looked at me and said: "Hey, you going to give me a hug for my birthday?" I said, "Why, you bet your sweet ass." I grabbed her up, hugged her ,and she kind of held on to me for a second. She turned right around and told the people she was with, "I am going to marry this man one day." I said, "Get me the hell out of here," and I took off. I was in Colorado working and lost touch with her. I got sick of Colorado and left. We met up again after that. We bumped into each other and started this courting thing and all the love stuff. We were married.

God had his things worked out that we didn't know; we couldn't see until now. I see where the paths intersect now. I prayed in 'Nam: I said,

"I'm new to all this, I don't know what's going on." His plan comes together, and I didn't see it. I came to this one road, and it's kind of rocky. I am thinking it's that one road that's in that valley of trials, torture, tribulations, and all that stuff—tests. I told God all I wanted was a little farmhouse. I said, "I don't want much, am not asking for a lot." I prayed and saw a little white house and a wagon road.

So we were married, and she got pregnant pretty quick right off the bat. Six months later, the Lord gave us this house. It was an eight-room farmhouse. Now it has 11. I ended up buying 37 acres of land for $29,500. I am on a ranch right here at No Business Mountain, and it's beautiful. God, it's beautiful. I love it. He made it possible that I could get this house, and, I kid you not, this house is what I saw in Vietnam. I saw it.

Agent Orange. I thought they were doing us a favor. I thought to myself when they were spraying this stuff that they were doing us a favor. I thought: "Well, bless their dang hearts. They are spraying us to kill those fucking mosquitoes."

I've had shit taken off my face. They say it was cancerous, but they hope to God they got all of it. You mention Agent Orange in Virginia, and they say, "What? Who?" They don't care about us.

I have a boy and a girl, and they both have spina bifida. I am trying to get them into the system somehow because they need to be taken care of. They could probably get some help or something

I went for this psyche evaluation. And it was a requirement that I go to these Vietnam veteran meetings and groups. I got there the first night and I listened to them. I introduced myself and all of them was talking about war stories—who's done this, done that, and all of that. Anyway, this one dude, he said that he had a Puff fly straight over his position and it opened up on the outer perimeter. I said, "That actually happened to me but I didn't like it." I looked over at him. I said, "Worst thing about that was those hot links and brass falling on you." He looked at me and he said, "What are you talking about?" I said, "You didn't feel the brass and the hot links falling on you? You said it flew right on top of your position." He said, "Lord, no." I got up, and they said when you leave a seat at one of these meetings it's an act of aggression. That is their protocol.

I more or less got into his face and told him he was a lying son of a bitch. I told him, "You are living on somebody's story. because I have been there personally, buddy, and I know for a fact that if machine gun belt links and brass fall on you don't like it."

It burns the shit out of you. So they told me to leave. I said, "I'm gonna leave alright and I ain't coming back, either." My shrink told me what you need to do is go home and just stay away from people who aggravate you. He said don't go around them. That's exactly what I do.

Vietnam did not change my belief in God. I was saved on May 10, 1960. God was all I could rely on. My favorite is Psalm 91, God's promise. He'll take care of us no matter what's going on. We can put our trust in him, our faith—it's his duty. Psalm 91 helped me internally and mentally to cope with some of the shit I've seen and done myself. I told my daddy that I had to carry it inside me. I did not tell one person, but it's about 27 years of carrying it. When they sent the chaplains out to the field, you had the Catholics, and you had the Protestants. I went to both services. I try not to offend anyone's religion. I know how I was raised and I'm not forcing mine on anybody, but I will tell you about it if you ask.

I have been to the Wall one time. I'll probably never go back. I went with the VA's PTSD program; they said it would bring closure and healing. Bull! When I saw my guys' names there it brought me back to the last time I saw them. That's why their names are there. How I longed to be with them. People look at you like a hero. I am just a regular, ordinary person. Just an ordinary Joe Blow, who has a beard and a cape. There is nothing special about me: I'm not an elected official, nothing like it. People walk up to me like a hero. I got my vest and wear my Bronze Star, my Purple Heart. But I am no different from anybody else.

I was woodcarving here one night. I couldn't sleep because I was worried about my wife, because of her open-heart surgery. I was woodcarving here in my kitchen and I looked up and, lo and behold, I don't know why, wasn't even thinking about nothing, but I was looking directly in the girl's eyes that spit in my face at Fort Lewis, Washington. I mean, I was looking directly into her eyes. I could see her face, I saw her eyes, the color of her eyes, I saw everything. I could remember it just like it was yesterday. I never even thought about this in all these years. Some things never go away.

The war took a 17-year-old boy and turned him into a non-caring, non-feeling killer. My experiences taught me a lot, and that is why I am the man I am today. I can take a lot. Touch my wife, kids, dog, or my weapons and have your funeral arrangements in order.

The war was useless! But, I'd do it all over again if my country needed me. It turned me into a man, quickly.

When I returned to civilian life, my soul was wounded and openly bleeding.

Specialist 5th Class James M. (Jim) Storey

268th TC Detachment, B/4/77 Aerial Artillery Battalion, 101st Airborne Division

My "Storey Story" covers almost 50 years. I sound like three different people and in a way I think that I have lived three different lives:

The person that I was before Vietnam;

The person that I was after Vietnam until about three years ago;

The person that I am becoming today.

I was drafted in the Army in 1967 for a period of two years. I was married only two months when I received my draft notice. I had just turned 19 years old. I couldn't believe it. I decided to enlist for an additional year in order to avoid being assigned to the infantry. They offered me Officer Candidate School but I decided against that. After basic I was trained as an aircraft airframe repairman and welder, MOS 68G30. I worked on Cobra attack helicopters. I was not surprised when I received my orders for Vietnam but I was not very happy about it.

My outfit was formed at Fort Bragg and we went to Vietnam as a complete unit in the spring of 1969. I flew over on an Air Force C-141 transport. It was a long, long flight. We landed in Da Nang and were immediately hit with a rocket attack. They hit the oil pipeline and I thought that I would never survive my year there!

We were part of the 101st Airborne Division. We "Screaming Eagles" were based at Camp Eagle near Phu Bai and the DMZ. We lived in a constant state of confusion, since I saw us under attack as many as four times in one night and I also saw us go two months and not be attacked at all. I patched bullet holes in Cobra attack helicopters in the daytime and dodged rockets at night.

Most of my flashbacks were caused by my own "night from Hell." From a sound sleep I was awakened to the night sky brighter than day, and thousands of explosions going off. I rolled out of my bed, unable to move, and was sure my days were over. By this time, we were seasoned combat veterans, but had never seen anything like this. We later learned that a single 122mm rocket had hit our ammo dump and set off over 2,000 of our own rockets.

After what seemed like an eternity, but was only a few moments, I regained my composure and half-ran and half-crawled to a nearby bunker. Everyone in the bunker was in disbelief of what was going on. Then one of my men came in the bunker in an absolute panic. I'll never forget his voice as he screamed, "There is a rocket in our hooch and it's on fire. It is still spinning and is going to cook off and explode. There are still guys in there and they are going to be killed!" God called on me and I had to stand up. I left the bunker and returned to our hooch. The rocket was still spinning and red hot! It was one of our flechettes that contained thousands of small nails with little wings like an arrow. If it cooks off, I'll be a pin cushion. Something deep inside me made me wrap the rocket in a towel and start out the door with it. It was so hot that the towel that I had wrapped around it caught fire, but I made it to a ditch and away from harming my friends. One of the dumbest things I have done in my life. I later was awarded a Bronze Star, but my reward was that none of my men were hurt.

That night we lost our hangar and ten of our 12 Cobras. Our unit was like a warrior with no sword. The fuel point caught on fire, and there were explosions that lasted into the next day. When dawn finally arrived, you could not walk without stepping on debris.

God was with us that frightening night, for in spite of all that damage and destruction, we did not lose a single man.

Then the suicides began.

Not long before I was due to return stateside, a good friend and I both received "Dear Johns" from our wives. We were both devastated.

I chose the light side, got the divorce, and lived with that scar. My friend chose the dark side and blew his head off with his M-16. I still miss my friend and always will.

Just a few days later I received word that my stateside cousin had taken her own life. She was a beautiful woman, well educated, but troubled. I still miss my cousin.

Not long after I returned stateside, we found a 14-old-girl, who was a friend and babysat for my girlfriend, near death from an overdose of sleeping pills. That ride to the hospital was the wildest one I have ever taken. I was in the back seat, shaking the girl to keep her awake, while my friend (another Vietnam vet) drove. The police tried to escort us but they didn't have a chance of keeping up with that muscle car 442 Oldsmobile. We saved her life that night, but a couple of months later she succeeded in taking her life with her mother's pistol. I still miss this young girl.

Coming Home

I extended my tour in Vietnam so I could get an early discharge when I came back to the States. If I didn't extend my tour I would have to spend another eight months in the Army stateside. I didn't want that. By extending my tour I was able to be discharged five months before my scheduled discharge date. It was worth it to me to stay in Vietnam a few more months and be out when I came home.

Unlike so many flights home, my flight home was totally silent. No celebration or cheering, just total silence. I flew into Seattle, Washington, and then on to Fort Lewis for my final discharge. We did not see any protesters or experience any hostile reception. When we arrived at Fort Lewis they told us we had to get a haircut or we wouldn't be released. A bunch of us refused. We had had enough of the Army; besides, what were they going to do? Send us to Vietnam? They finally let us go, anyway. They actually paid very little attention to us. It was almost like, "Don't let the door hit you on the ass on the way out."

I never had trouble nor was mistreated on the way home. Actually, on my flight from Seattle to Atlanta, a nun insisted that I take her seat in first class to honor my service. It was like I was in an unreal world, completely different than I had been in the 15 months in Vietnam.

I had been divorced while in Vietnam so I went to my parents' house right away where they had a room waiting for me. My family welcomed me home very warmly. They were confused by my not wanting to talk about my Vietnam experience.

It seemed like everything had changed while I was in Vietnam. The only people that I felt comfortable with were other Vietnam vets. I just couldn't communicate with non-vets. Most of my old friends were gone. Most of them had gone in service or left for jobs far away. I felt disconnected from everyone. In a way I still do today.

Home

When I returned to civilian life, my soul was wounded and openly bleeding. I hated politicians and the people who had sent me to a foreign land. Then these same people didn't support the very troops who were fighting for someone's freedom. I tore up my dress greens, threw away my Bronze Star and other medals, and began openly opposing the war. Today I am ashamed of these and other things that I did, and hope people can forgive me.

When I returned stateside I applied to and was accepted at Georgia Tech. Not bad for a southern boy who didn't know that many three-syllable words! I didn't make it through the first semester. I was too scarred to concentrate on things like chemistry and calculus.

A short time later I got a good job with a leading home improvement company as a traveling salesman. I was highly successful, set new district sales records, but continued to spiral downward. After 18 months, I quit the best job I have ever had and went to Colorado to ski. I was totally disconnected from life. I became an adrenalin junkie. Fast cars, fast women, and fast skiing were my outlet. These were the only ways that I could feel any emotion. I still deal with this today.

My time at Vail probably saved my life. My brother and I were full-time ski bums. We skied 125 days that winter and managed to not work a single day. It was an experience I wouldn't trade anything for. We lived in Avon before Beaver Creek was started. There was no one there except us poor ski bums and thousands of sheep.

Toward the end of the year I became depressed and started my downward spiral again. I remember one cold night, stupid drunk and walking down the railroad tracks that ran through Avon. I barely remember laying down on the tracks waiting for the train to end my misery.

Somehow God was not through with me, for he picked me up from the tracks that night. I drifted around for about two more years and finally made a turn from the dark side. I guess I had finally bottomed out.

I finally went to the VA in 1976. I talked to a counselor there, and after just two sessions she helped me realize my pain. The doctors told me I was suffering with survivor guilt and posttraumatic stress (they didn't call it PTSD then).

I was in Colorado in 1975 when Saigon fell. I was totally devastated. I felt like we had been betrayed by the politicians. All those lives, all that time, and all that money was flushed down the toilet.

I have dealt with liver abnormalities and skin problems since my return from Vietnam. It took my liver 15 years to recover and not to raise red flags in my blood tests. When I mentioned Agent Orange, they sent me to the psychiatrist. I went twice and never returned.

My family and friends said I was a different person when I returned from Vietnam. I would alternate from anger to being silent and withdrawn.

I was not very spiritual before Vietnam and when I returned I called myself an agnostic for many years. Now I have been reunited with my faith and now am active in my church. It took a long time.

Looking back, I wish I had not spent 40 years denying my problems. I am now healing by helping others who are suffering the way that I did. My nightmares and flashbacks are much fewer than they once were.

The next 20 years of my life were much better, more productive. I had two different businesses, got married to my wonderful wife, and seemed to be growing old with grace.

There were still rocks in the road. I can remember on Memorial Day in church, when they were honoring our fallen warriors, that I had a bad flashback. Once again I could see and feel the rockets' red glare, the bombs bursting in air, and feel myself descending into the madness of hell. I had to leave church that morning. I never explained it to anyone and they probably thought I was just emotional.

Then the first time I saw the Grand Canyon, God found me again. I remember looking over the canyon and realizing there really is a God, and he has been watching over me. My days as an agnostic were over.

When I retired, I fulfilled my lifelong dream to return to Colorado. I thought I was coming to ski, but later realized I came back to Colorado to be closer to God. I realize the value of wilderness therapy by feeling closer to God on the ski slope than in church.

In the spring of 2013 I read the book 'War and Recovery" by Vietnam survivor Dave Roever. I realized, after denying my problems and having flashbacks and nightmares for over 40 years, that the time had come to do something.

For three years I attended sessions given by the Roever Foundation at their Eagles Summit ranches in Colorado and Texas. I went on to pass the test given by The School for Military Ministry, founded by Dave Roever. Over this three-year period, I was able to change from a suicidal veteran with no hope into a "Screaming Eagle" wanting to help other vets who were suffering like I was.

Today I use my personal appearances to meet vets and work with them on a one-to-one basis. I listen to their stories, tell them they are not alone, and assure them there are better days ahead. My "Storey Story" helps "Welcome Home" those who had given up hope.

My main interest is suicide prevention, having lost vet friends and family members to this tragedy. Twenty-two veterans a day commit suicide, and this is totally unacceptable. If I can reach and save just one, it is all worthwhile.

Chapter 3 - Leaving Vietnam

The standard tour for the Army was 365 days. Soldiers started counting down the tour the day they arrived. When they had less than 100 days left they were a "two-digit midget." When they had a month to go they were "short." The Donut Dollies would pass out short-timer calendars, and the troops would mark off the days until they boarded that Freedom Bird and went home.

About a week before their DEROS date (Date Eligible for Return from Overseas), the rotating soldier would clear post, check out at battalion, turn in his gear, say goodbye to his buddies, and head off to Cam Ranh Bay, Da Nang, or Bien Hoa to wait for his or her Freedom Bird.

The trips home for Vietnam veterans were as varied as their tours. The only common thread was the happiness and thrill of leaving Vietnam alive and going home.

I came home from Vietnam twice. I extended my tour for an additional six months (I actually spent one year, six months, and 28 days in Vietnam) and after 13 months I was given a free 30-day leave. The morning I left my company there were very few people around. They were all working. So, I packed my bag, looked around at my hooch (I would be back in 30 days or so) and headed out to the flight line to catch a hop to Bien Hoa. It felt weird to be leaving Phu Loi after a year, unsettling. It was November 1971.

I boarded a Huey slick for the flight to Bien Hoa. As we left the perimeter and headed out to open country I admired the beautiful, lush scenery. I just sat there and took it all in: the small villages, rice paddies, and bomb-scarred landscape.

After processing at Bien Hoa I took a bus to Tan Son Nhut where I waited for my flight home. It was hot, but I was used to that. I was comfortable in the warm weather of Vietnam. I changed into my khakis and kept my field jacket out, as I knew it would be cold in the United States.

My route took me to Japan, Alaska, and ultimately Travis Air Force Base. I flew on Flying Tiger Airlines. It was the same route I flew over to Vietnam and the same carrier. I don't remember anyone cheering when we took off. I remember the silence. It was deathly silent. After about a half-hour after takeoff, you could hear chatter in the plane. We were on our way, and it was not a dream.

We landed at Travis Air Force Base in California on November 22, 1971. We took Army buses from Travis to Oakland Army Depot. As we entered Oakland we were met by protesters, who threw fruit, garbage, and God knows what else at our bus. They held signs calling us baby killers and gave us the royal finger. This shouldn't have surprised me, as I knew about the protests when I left for Vietnam in October 1970. But the reality of it hit me hard. What struck me was not the protests but the venomous way they carried it out. There was a lot of hate there and it was directed at us. Welcome home.

Some vets have told me they received a steak dinner and new uniforms when they arrived at Oakland Army Depot. Not us. We were processed and bused to San Francisco to fly home. I was in my khakis and field jacket, and it was cold!

The flight was relatively uneventful. A couple men started grilling me about the racial situation in Vietnam. That was it. Otherwise, I was pretty much ignored. My parents and my brother picked me up at Chicago's O'Hare Airport and drove me home. I was freezing. Everything looked so foreign. It was late evening and all the lights captivated me. It was like a dream.

I tried to relax and settle in but couldn't. I was edgy and uncomfortable. I felt awkward around old friends. Their lives were different now and mine was, too. I turned 20 while I was home on leave; I still could not buy a beer. At bars I would show my orders proving that I was home on leave from Vietnam but the bartenders didn't give a damn. Not one of them. They would throw me out and threaten to call the police. I can't explain how angry that made me. After being home a couple weeks I started to get anxious to go back to Vietnam. I just felt completely out of place.

I was supposed to report to Fort Dix, New Jersey, (I can't explain Army logic on that one) on December 22 for the trip back to Vietnam. Three days before Christmas? I doubt it. I flew to Fort Dix after Christmas, on December 26. The captain who checked me in looked at my orders, then looked at me and said, "You are AWOL, soldier." I looked at him and said, "What are you going to do, Captain? Send me to Vietnam?" He leaned back and said, "You have an attitude problem, son." I said "Listen, Captain, I spent last Christmas in Vietnam, and I am going back. There was no way in hell I was going back to Vietnam three days before Christmas." He stared at me a couple of minutes then looked at my orders. He changed the date on his stamp to December 22, stamped my orders, and checked me in. He said, "I spent a Christmas there, too. Your flight leaves on January 1. You'd better be on it." I saluted and said, "Yes, Sir, Captain." I left the orderly room and went back to the airport and flew back home. I flew back to Fort Dix on New Year's Eve and was there for my flight to Vietnam on January 1. I saw the captain, saluted, shook his hand, and thanked him. He wished me luck, and I was on my way.

I landed at Bien Hoa, midday. Hot, humid, stinky. I looked out at the sprawling base and felt I was home.

Army Vietnam Vet: That was one of the greatest days, when you caught the Freedom Bird. You knew you were safe when you got on that plane. But then when you got home they were calling you a baby killer.

Marine Vietnam Vet: I returned to the States in November 1968. My tour in Vietnam was over with MAG 16, from Khe Sanh in January 1968 to Da Nang at the end. It had turned a well-trained young boy into a man with many issues and doubts. Be that as it may, my day of redemption was at hand. On the runway at Da Nang was my silver bird, compliments of Uncle Sam and World Airways. I would say 90 percent of us were on board and were in our seats. From my seat, I could see strange happenings going on outside, although they should have been very familiar. Everybody was running and diving for cover—the light came on! The airfield was under rocket and/or mortar attack. As I grasped that reality, the pilot came on, yelling, "Everyone to their seats!" The plane door closed and the next thing I knew we were airborne, circling the airfield. We could see the attack taking place! The stewardesses were picking themselves up from various parts of the plane as we circled twice and headed over the Pacific to the Philippines. The rest of the trip was uneventful until we reached the States, where we were greeted in San Francisco with protesters.

To this day, I wonder how many died in that attack in Da Nang and never made it home. So close and yet so far. To this day, Vietnam haunts my memory as well as my dreams.

Army Vietnam Vet: Thinking, "Holy shit, I made it!" Butterflies in my stomach! What is the first thing I want to do when I'm home? Then the guilt sets in; why me and not my buddies?

Army Vietnam Vet: I came home on a medevac, but the flight nurses in fatigues were hot, especially the brunette that hand-fed me the fried chicken, because I had little use of my arms. I'm good now, but it is a great memory.

Army Vietnam Vet: Awwww, yes, the day I left to come home it was pouring rain. One hundred and ninety-nine G.I.s were told, "There's your ride, run for it!" I was probably number 180 getting on. I was soaking wet. I fell asleep right after take-off and woke up with a pillow under my head and a blanket over me. A stewardess was standing there and smiling at me. I was grateful to her: that was the first sign of kindness I received after I left 'Nam.

Army Vietnam Vet: What I remember is that we had to wait to leave Bien Hoa because the flight line was being attacked. After taking off, it was a long, 18-hour flight to Oakland. Got meals a couple of times. Oakland was the worst as we had to sleep under bright lights, but we were home. It was Easter weekend, and we had dinner on Fisherman's Wharf.

Donut Dollie Vietnam Vet: That rapid (though the flight was long) change from in-country to The World made a big difference. I believe that even Korean War vets came home mostly by ship. That time on a ship during WW II or Korea coming from Europe or Asia helped in the decompression and adjustment. Vietnam vets did not have that. Add in the factor that there were the protesters and no warm greetings as in WW II, it all made a difference in our homecoming.

Army Vietnam Vet: I can remember getting on that Freedom Bird, in khakis. But I didn't care at the time what I was wearing—just let me on. As we were taking off, rockets started hitting the end of the runway, but nothing hit the plane. We got airborne with no trouble. I respected those Freedom Bird pilots with the same respect I had for the chopper pilots.

Army Vietnam Vet: Over there, we felt that we were doing something important, and then, all of a sudden, it was over. Coming home—we all wanted to come home, but we didn't want to leave.

Army Vietnam Vet: I still can't explain it but when I got in the Jeep to ride to the airstrip and looked back at the company CP, I cried all the way to the airfield. The two other guys in the Jeep just looked at me in disbelief. Maybe and hopefully they experienced the same emotion when it was their turn to leave Hell. They didn't say anything.

Chapter 4 - Navy Stories

The Navy had a very complex role in the Vietnam War. The variety of duties they performed was unequaled by any of the other branches.

The "Blue Water Navy" remained offshore and supported ground troops with shore bombardments and air strikes. Cruisers, destroyers, and a battleship bombarded targets in North and South Vietnam. Air assets from aircraft carriers provided close air support for the infantry in the south and struck targets in North Vietnam repeatedly throughout the war.

In addition to the combat vessels, there were many support ships that kept the Blue Water Navy supplied with ammunition, food, and fuel. The Navy also patrolled the coast of South Vietnam, searching for infiltrating enemy troops and supply ships.

The Navy was very active within the geographical boundaries of Vietnam. They operated on the land and in the waterways of South Vietnam. The "Brown Water Navy" utilized river patrol boats and fast patrol boats (swift boats) on the waterways and rivers of South Vietnam. They heavily supported the Army's 9th Infantry Division in the Delta of IV Corps, and the Marines in I Corps.

They were supported by LSTs, LCMs, and various supply ships as well as by land-based logistical support sailors. These ships also supplied Marines traveling the hazardous rivers in I Corps.

The Coast Guard in Vietnam served under the Navy. Approximately 8,000 Coast Guardsmen served in Vietnam during the war. Seven lost their lives there and 60 were wounded.[1]

The Navy handled mass logistical operations in Vietnam, including harbor operations and the huge influx of material to supply the ever-increasing war effort. At the peak of the war effort in 1968 there were approximately 38,000 Navy personnel in Vietnam. This number decreased to approximately 17,000 by 1970.

The Seawolves were Navy helicopter gunships that supported the Brown Water Navy. They were stationed at various firebases in III and IV Corps.

The Navy Seabees were very active in Vietnam. They constructed Marine bases, fire support bases, roads, structures, etc. They were the construction force of the Navy and Marine Corps.

The Navy supplied all the corpsmen to the Marines for battlefield medical care as well as hospital staff, including doctors and nurses.

The Navy and Coast Guard suffered 2,565 deaths, 4,178 wounded requiring hospital care, and 5,898 wounded not needing hospital care.[2] One hundred seventy-nine personnel were taken as prisoners of war.

Twenty-one aircraft carriers conducted 86 war cruises and operated 9,178 total days on the line in the Gulf of Tonkin. Five hundred thirty aircraft were lost in combat and 329 more to operational causes.

1. Coast Guard History, United States Coast Guard

2. Casualty figures from Defense Casualty Analysis System (DCAS)

BM2 Bob Johnston

Naval Support Activity, Da Nang, US Navy

I joined the US Navy in September 1964. I was the only son on a family farm and the oldest child. I wanted to get away and see some of the world. I'm from Slate Lick, Pennsylvania, about 65 miles northeast of Pittsburgh. I wanted to join right out of high school, but my mother talked the recruiter into delaying the entry until the fall. I was 18 and knew nothing about the war because it wasn't really going on yet; at least, there wasn't much in the news about it.

I went through basic in San Diego, California. At the time all of the schools I wanted were full; the fleet was full and, although I wanted subs, I failed the tank test due to bad sinuses. I wound up being assigned to one crap job after another: mess cooking, night cooking, security, and a bunch of BS things that had me more pissed than anything else. You couldn't make rate and I was broke in a "fleet" city with no future.

A friend told me one day that they were looking for volunteers to form a group to go to Vietnam. There was extra pay in it and, being Navy, how bad could it be? I had no girlfriend and wasn't married, so I volunteered.

I completed SERE training (Survival, Evasion, Resistance and Escape training) and weapons training. It was two weeks with a group of the 5th Marines at Camp Pendleton. After training we just sat around for a month, 427 of us. In the end, some of the others in the group left in jet transports. My group, the last one, got a C-130 and a one-week tour of the Pacific . . . every island that had an airfield!

I had no MOS and no training except basic weapons. A lot of us were assigned to general support for the Naval Support Activity, Da Nang (the Naval Support Activity in Da Nang provided support for the Navy and the Marines in that area.) I was assigned to mess cooking, again! I got out of that by pissing off every petty officer (NCO) and officer I could.

After I got out of being a cook, I was assigned to lighterage where I hauled cargo and anything else that had to go somewhere. At first, I just moved cargo from ships in the harbor to a sand ramp up the river. Then it was troops and vehicles. Finally, it branched out to what was called "coastal operations." I was started hauling cargo and troops up and down the coast, and then my boat was assigned to places like Tan My and Cua Viet (bases established at the mouth of the two rivers) for supply/patrol/interdiction. I spent some time on LCM 6s, LCM 8s and LCU/YFUs.

Our main job was to haul any cargo from ships we met sea-side or at a ramp at the mouth of the rivers to offloading/loading areas upriver. The cargo was everything from palletized C-Rations, other food, munitions, vehicles, people, and fuel. Our task was to run the river and deliver the goods, no matter what, as many loads a day as we could fit in.

I was trained as a boat handler after going aboard a Mike 8 (Landing craft, mechanized/medium 8) as a seaman. The ones we had were 72' long and 22' wide and mostly well deck for cargo. They had four supercharged GMC 6-71 diesels arranged in "banks" of two, each "bank" driving a three-blade prop through an Allison automatic transfer box. The boats weighed around 50 tons and could haul 60+ tons of cargo in five feet of water. They had enough power to dig their own channel in shallow water. Top speed (loaded) was about 13 knots or maybe 15-18 MPH. The boats were delivered to us unarmed and unarmored: it was up to us to come up with weapons for them.

They were very short of craft commanders (boat drivers) so even before I was an E-4 I was given command of my very own 50-ton, 74-foot Mike 8. I learned well and could run one of them without a problem. So well that I soon found myself taking the boat on "coastal operations" up the coast to the wonderful Cua Viet River. That was in the early summer of 1966 in support of Operation Hastings. That was the beginning of three tours in hell.

Coming Home

On July 13, 1968, a runner was sent to find me at the ramp in Cua Viet. Our NCOIC had onion skin orders for me that I was going home. I was told to get off the boat and get my few things together and hide in a bunker; my relief was on the way up from Da Nang and he would be there the next morning on a U-Boat (U-Boat was a general way of talking about an LCU or YFU. Larger than our boats, slower, and hauled a larger load. They, along with LSTs, got the cargo to us then we'd haul it up the rivers.)

I spent that day and the night in the bunker. It was crowded because we took a few hours of incoming that night. I'd been up there for nine straight months. The next morning my relief was on the boat, but it didn't stop at the ramp. On the way to Dong Ha, they took direct fire from the riverbank, and he was among the first KIA.

I was then told just to get out any way I could. There was a Huey on the ramp, so I grabbed all my gear and begged them for a ride back to Da Nang. That was probably the worst damned hour-and-a-half of my life—I just knew we wouldn't make it, and I'd wind up dead after all.

Everything I owned at that point amounted to a faded Hawaiian print shirt, another in my field pack, along with a second pair of cut-off green pants, a few grenades, a basic field toilet kit, shower shoes, my cut-down M1 carbine, along with a 45 and extra mags for each of them. I also had my ID, a few pictures, dog tags, my steel pot, a sweat towel, and a worn-out flack vest. When I checked in at the White Elephant, everything was taken away except for my orders, ID, some MPC, and the clothes. I had no time to draw a new uniform because my flight left the next morning.

I went straight to the airport and camped out in a stall with an old man selling gedunk just outside of the gate. Everyone avoided me, probably because no one knew who the hell I was. The next morning I had breakfast with the old man, had some coffee, and drank beer until the plane was ready to go. I had the field pack over one shoulder and the orders in one hand. I went through the gate and boarded. Yes, there were some questions about my "uniform" but I think my "look" told them, "You fuck with me now, and you're dead."

There wasn't a sound on the plane when we taxied out and turned around. The flight was all Marines except for me.

When the wheels broke ground, we all cheered so loud that we almost scared the stews out of their pantyhose! We cheered, we clapped, and we cried; a lot of us were crying. I popped open a beer and shared another one with the Marine sitting beside me. I guess I was feeling total elation.

We refueled in Japan and Hawaii and finally landed at LAX. I don't even remember how I got to Long Beach, but that's where I reported to be processed out. The first thing I did was pick up a few pair of civvies to get off base. I let my hair grow and spent a month just staying drunk, either on the base or in town, but never in the area where I'd run into any service people.

Since I was being processed out no one wanted anything to do with me. I didn't talk too much to anyone and spent a lot of time trying to get in touch with anyone in my family. I was exactly 22 years and one day old.

By the time I was "out" I'd found out that the family home had burned to the ground on July 5th. My family was scattered to the wind.

I had two younger sisters; neither they nor my mother knew exactly where I was in-country. Had my mother known where I was she would have probably gone nuts. As a result, no one was really concerned about the fact that I was in Vietnam, and it was no big deal that I was home. My father had the details, but he didn't give any out. As a result, they all thought I was in the rear with the gear.

Emotionally, I was pretty screwed up. My life, before July 5, (when the house burned down) didn't exist. Everything I owned was in the bag I carried when I got off the plane in Pittsburgh. By the time I got to the farm the hole was filled and the grass was planted.

I instituted my own policy of just telling people that I'd been in the Navy. That's all: I didn't talk about Vietnam. It was less complicated that way.

I wasn't comfortable back in the States. I wanted to be there, but it felt like I was on a different planet with a bunch of strangers. That goes for family and friends. Most of the friends I'd had in school had moved on or were going to college; some were dodging the draft. I simply didn't fit in anywhere.

There was just no one to talk to about anything. I was on my own.

Home

The disconnect only got worse over the years. The only person that I ever became close to again was my mother. Before she died I let her read some of the chapter proofs of a book I wrote. Those were some pretty tearful sessions. Two marriages and a bunch of relationships went on the rocks: my children broke off all communication about 15 years ago. I guess I expect too much of people.

I had a series of jobs, a few businesses (some successful, some not). The biggest problems in the jobs seemed to be that I like to get the job done and done right, on time and without drama. After the things I went through—as we all did—normal life is easy, no need to complicate it or add drama. You also expect others to do their job and keep their word about things. This is where problems always seemed to come up.

I totally tuned out anything about Vietnam. When the announcement came that Saigon fell, I was sitting in the cockpit of a 50' yacht in St. Croix getting pleasantly pissed. I was about to take a party out for a week in the sun . . . I made myself another drink.

I ended up moving to Vietnam and opening Bob's Café American and Motorbike Rental. I always liked the country and the common, everyday people here. Since I spent most of my tours on operations on the rivers in the boonies, that's the kind of people we dealt with daily. I had a good dose of the guilt that I think we all felt and feel about the war: it's overrated and misplaced. During the first year of my return, I lost all the, "I'm going to do something to change things for the better" crap, and settled into reality.

I came back to Vietnam to teach and relax. I was tired of what I saw my country turning into and didn't like it. I wanted my life to be easier, simpler, less complicated. When the school I was working for closed, I just decided to stay. By then I'd become part of the community and had local friends. I live in Tuy Hoa because there are no memories here and nothing to remind me of the war.

I have a hefty dose of PTSD. Living here has done a lot to relieve some of the symptoms; however, some are magnified. I've learned to deal with it. I get no compensation from the VA due to the way my DD 214 was filled out. None of the boats on coastal ops. and none of the crew were ever shown serving up those rivers. Mine shows three tours as "NSA Da Nang" and there's no way to change it. I was called a liar by the VA and told I was just looking for welfare.

While I was down at the Liberty Bridge, we were under fire and attacked several times. There was always direct and indirect fire at our boat and the Marine compound there. In Cua Viet, we lived with this on both ends of the river and during the transits (we could make up to six runs a day). We were subjected to indirect fire at night. Most nights we slept in the bunkers due to rocket, mortar, and artillery fire. Imagine being on a boat with 50 tons of 155 or 175 rounds, pallets of powder and fuses for the rounds, and then going up an unsecured river (the Marines did the best they could) 12 miles at 15 MPH, knowing that a kid with a BB gun could punch your ticket?

We ate C-Rations or what we could trade with the locals for; there was no chow hall. We supplied everyone else but couldn't get parts for the boats; we couldn't even get uniforms. We would steal canned food from broken loads to have something better to eat at times. We survived.

People were looking at me like I was a murderer.

EOHCA James Stroud

Mobile Construction Bn. 71, U.S. Navy

I wanted to join the Marines, but my mom said why not see about the Seabees, as I liked equipment, so in 1962 at 17 I joined the Reserves. I dropped out of school then went on active service. I served for six years.

I went to Vietnam in 1967 with MCB-71. We flew to Vietnam in a C-141 with passenger seats. We landed in Chu Lai where we unloaded and boarded our horse trailer for the ride to Camp Shields. We went into our huts and started unpacking, when the siren went off. Myself and the other guys in the hut were looking at each other when we realized we were not in Camp Lejeune anymore! Rockets started coming in, and we grabbed our rifles, vests, and helmets and jumped into our pits outside our huts. I was thinking that this was the VC's way of saying "Welcome to Vietnam."

Everything was going fine. I was assigned to the asphalt plant where I operated the north west 6 crane, loading the hopper. Every morning, six-and-a-half days a week, I would go out to the plant and go over the crane, making sure there was nothing under it that might set off a charge. For almost three months we did this, then the asphalt plant was shut down. I then went to the motor pool where I was put on the night shift.

It was bad enough going on the night shift and working in the middle of the woods, but I also received a letter from my girlfriend telling me that she was pregnant and was dumping me. This was the beginning of my downfall.

I was assigned to a crew of 25 men working about ten miles away from our base. We had no security. We were our own security. We were building helo pads for the Army. I just remember how dark it was behind me while operating a dozer. I started smoking weed and getting high. Life sucked. I`m thousands of miles away from home, my girlfriend just dumped me, and there was nothing I could do. After about a week or so I was getting real bad, and I started worrying about snipers more and more. One night we were running dozers when we came under a

Condition Red. The VC was in the woods, coming our way. I was a senior E-4 and was assigned three other men. We went into a trench, where we took our position. I took the safety off my rifle and told the other guys to make sure their rifles were not on auto. (The next night I received a field promotion to E-5.) Now, what I thought was help was the start of my nightmares. The choppers started dropping parachute flares and the ground around us lit up like it was broad daylight. Then the choppers started shooting, but the rounds were all around us kicking up the dirt. I couldn`t believe it! Were we going to be hit by friendly fire? Our lieutenant was able to call and get these guys off us, and all of a sudden darkness fell all around us. It became deathly quiet. After about ten minutes, the lieutenant had us all get back in the trucks and return to our camp.

Coming Home

Now I didn`t care at all what happened to me—just let me get high. Six months on nights and I became a mess. I started getting into trouble, and when we finally returned to Davisville, Rhode Island, nine months later, I was drinking heavily and doing stupid shit with drugs.

I was able to go home and was treated like dirt. People were looking at me like I was a murderer, calling me a baby killer, and spitting at me. I`m trying to figure out what the hell did I do that people were pissed off at me. Then after watching the news a few times and seeing how we were being portrayed as a bunch of murderers, I started understanding. The people were being fed bullshit. Why aren`t the news reporters showing the snipers we had to put up with at night, or the booby traps guys were setting off?

I reported back, but I didn't care anymore. A lieutenant came down from the quarterdeck and walked into the barracks where I was smoking weed and drinking. I know someone called him and let him know what I was doing. He walked into the barracks, saw me, and said I was under arrest and the MPs were on the way.

Now I had had it, I was so high I swung at him then ran out of the barracks and stole his military P/U truck and took off. The next thing I knew I was in Florida, out of money, out of gas, and hungry as hell. A

Florida state trooper stopped and asked if everything was ok. I told him what I did, and said, "I think I`m in trouble." He notified the military base, and two SPs showed up. I was then taken to the base where I had to get a physical as they can't put you in the brig without one.

After that, the SP took me to the chow hall where they were serving lunch. We went to the front of the line where the SP was going to show me he`s boss. He told me to take a tray and silverware. I grabbed the knife: that got me a club across my handcuffs. He told me, just a spoon. On the chow line, they were serving chicken. The guy put a leg and a wing on my tray. I looked at him and asked if I could have another leg. He looked at my cuffs—prisoners are marched to this chow hall, but never in cuffs. The guys in the chow hall figured I must have killed

someone to be in cuffs—took the wing off my tray and put four more legs on my tray. The next guy put two scoops of potatoes on my tray. The SP asked me what I wanted to drink, and I said, "Coffee." That got me a shot to my back and in a loud enough voice he said: "You will say 'Sir'." Oh, yeah, he was going to show me who's boss. He grabbed a cup and filled it with water. He then walked me over to the brig table where I sat down and started eating.

Now, I don`t know what happened next. I think it was a little bit of me sobering up and coming down from my high, but there was one Marine guard giving this young Marine a hard time, poking him in the back and just abusing him. Well, I'd had it, so I took my tray and threw it at him. The SP came over and put his club around my neck and pulled me down onto the floor. He literally dragged me out of the chow hall that was as quiet as a morgue. He threw me into the back of the truck and took me to a holding cell where he asked me, "What the hell were you thinking?"

Make a long story short . . . I went to the brig. The next morning I was taken into a room where three Marines started screaming in my face as my head was being shaved and told, "I will say 'Sir'" before and after I`m asked a question. Now, things were going okay, I was behaving myself, but the third night I was in my cell waiting for them to open up and march us to the chow hall, the gate opened, and three Marines came in. I thought to myself, "This isn`t good," so, being a stupid Seabee, I charged them. That didn`t go too well. I got pounded to the ground, and the last thing I remember was the Marine who I threw the tray at said, "I got you," and stomped on my ribs, where he cracked two of them. I woke up in the hospital, where I stayed for a week. My left eye was closed, and my right eye was swollen, but I could see.

After a week, I was brought back to the brig where the sergeant informed me that as long as I didn`t say anything they would not add the charges that I attacked the guards. Which was right, I did go after them. For the next two weeks I was left alone and put on light duty while my two cracked ribs healed. I was taken back to Davisville, where I went into the brig in Newport, Rhode Island. Three weeks later I was facing a court martial. I was charged with striking an officer, stealing a truck, AWOL, plus other charges. I was sentenced to ten years in Portsmouth, New Hampshire. I served eight months in the brig and was released, and discharged with an undesirable discharge. After 40 years, my discharge was changed to honorable. I got out in October 1968. That's my story—not much action but a lot of stress which I did to myself and deserved whatever I got. I can`t blame anyone but myself.

Home

My return to the States was a nightmare for me. I just wish I could have handled things differently. Thoughts of suicide entered my mind, but I fought them off. I started driving over the road, which was like therapy for me. My dream job was to be a New York state trooper, which was thrown out the window due to my record, but I did okay with my life.

First, I drove buses for a small bus company, and then I started driving tractor trailers over the road. I was hauling chemicals in tankers. In May 1975, I suffered a serious chemical accident when a hose ruptured and blew 1,500 hundred gallons of formaldehyde on me. My safety goggles were blown off my face, and I was pushed into the tires of the truck until the hose was empty. I was left blinded, but the doctors worked on me and made me drink some stuff, as the formaldehyde had been forced down my throat. The vision in my right eye came back even though it burned. I could see, but my left eye was almost destroyed. It took almost three years for my vision to return, even though it's a little blurry. For three years, I tried to figure out what to do. I had no high school diploma and very few skills. I went back to college through the VA and in three years I received my high school diploma, my associate degree, and my BA. In 1978. I started driving gas tankers locally until I retired in 2001, due to having strokes.

I now have three grown boys and ten grandchildren. My wife and I have been married for 42 years. If it weren't for her, I probably would have been one of those guys who committed suicide. I really was in bad shape when I got out. There were times I kind of wished they had taken me to Portsmouth for ten years, but of course now I'm glad they didn't.

I am considered 90% disabled. I still have another claim with the VA I've been fighting for nine years, but I'm not giving up. This Agent Orange is slowly killing us.

CM3 Eugene C. Giunta

Mobile Construction Bn. 3 (MCB-3), U.S. Navy Seabees

I joined the Navy Seabees in 1966 and served until 1970. I had received my draft notice, and enlisted to have a choice of service branch and job type. I was trained as a construction mechanic of heavy equipment.

I received orders for Vietnam twice during my four years in the Navy. The first time when I received my orders I was overly apprehensive and unsure as to what I was in for. When I received orders for Vietnam a

second time I never thought I would make it back home alive, but, thankfully, Nixon began troop withdrawals in the summer of 1969 and we were, at the last minute, sent to Okinawa instead.

I was assigned to Mobile Construction Battalion 3 (MCB 3), which went over as a complete battalion to Da Nang in I Corps, the most northern sector of South Vietnam. We flew over in a poorly lit C-141, with seats facing aft. It had only one window, at the back door. When the back cargo door of the C-141 dropped open, allowing the heat, the smells, and sounds of Vietnam to enter my senses, I didn't quite understand that this was a life-altering trip. From that moment forward the naivety embracing my character systematically unraveled.

The eroding was subtle at first: M-16 rifle fire popping constantly around me and 105's and 175's blasting in the background. This became routine and I accepted it easily as a part of my environment. After all, I was a Seabee; we were there to support the Marine Corps by building buildings, creating airstrips in the jungles, and building roads and bridges to move the troops from point A to point B, and, if necessary, to defend against the enemy. As a heavy equipment mechanic, I was most anxious to work on the legions of equipment we would use; the war was of secondary concern to me most times, and at others a nuisance.

Then in early February 1969 Tet reared its ugly head. Our perimeter in Da Nang, at Camp Haskins South, had to be protected from the suicidal attacks of the NVA and the Vietcong. Due to Tet of 1968, our military leaders were now highly sensitive to the escalation in the frequency and type of assaults dispatched on us during this holiday season. So instead of following our normal routine of work, eat, and sleep, guard duty was added to our daily process. This meant sitting in a foxhole with an M-16 in hand, or in a 30-foot watchtower manning an M-60 or 50-cal. machine gun at night to guard our camp.

Coming Home

I was overjoyed when we were leaving Vietnam. Our return trip to the States was on a commercial jet, a 707, with real stewardesses and regular airline seats and hot meals, not box lunches. It was great. When we landed, I was anxious to get back behind the wheel of my 1965 Chevy Impala and cruise the streets of St. Louis.

The only noteworthy event was when we landed in the States we were greeted for a photo op. by Richard and Pat Nixon. They were out in southern California to inspect an oil spill and decided it would be to their benefit to shake hands with a bunch of Seabees who had just spent 20+ plus hours on a jet and wanted to go home on leave.

Much to our chagrin, we were kept on the plane for two extra hours to shore up security for Mr. and Mrs. Nixon. This made for a lot of unhappy Seabees, me included.

Aside from the pomp and circumstances of the President Nixon photo op., which clearly was done for his own personal self-interest, things were business as usual: just another Seabee battalion returning from a tour in Vietnam, no welcome home, no recognition for any of us.

I was very excited and relieved to be home; it was like a burden had been lifted off my shoulders. In a way, it felt like R&R, because I knew in six short months I'd be going back to Vietnam again: that's just what

Seabees did, a ten- to eleven-month tour and then back again for another one. The boys in Washington had figured out a way to screw us into 20-22 months instead of the 12 or 13 months that almost everyone else did.

Walking through the LAX airport in my dress blues, wearing my ribbons, which gave away the fact I'd spent time in Vietnam, made for some hostile stares and glares. When I got out in March 1970, I made up my mind to go back to college.

My folks picked me up at the airport. They were all smiles and extremely happy to see me. I went home right away, but I didn't stay there: I was out the door to see some of my old buddies after a short visit with my folks. I found out quickly that most of my friends were busy with their own lives and, while they were glad to see me, they were preoccupied.

I was comfortable at home, at least at first. I was caught up with friends and family. Coming from a large Italian family and living in an Italian ethnic community helped. There were plenty of folks who said hello, it was easy to get caught up in that. We talked about everything while intentionally ignoring the Vietnam conflict. Then after a week or so into my 30 days' leave, things settled down and the gravity of the prior 11 months began to set in. Sleeping became somewhat difficult at that point. I remedied that by going out and drinking and closing the bars most every night. That helped a bunch.

I'll never forget; one Monday morning the city ran the test on the tornado warning sirens. I was still in bed asleep when the sirens went off. I was halfway down the stairs before I realized I was safe at home in St. Louis and wasn't in the midst of a rocket and mortar attack by the VC and the NVA.

But, things were different. My best friends had either already graduated from college or had recently gotten married. Both circumstances made me feel like a fifth wheel when I was around them. We all had changed. My once easy-going and laid-back disposition was buried, and I now carried a slight edge to my persona in both my actions and dialogues. I'm sure I put some of my friends off by that. My behaviors made interactions somewhat uncomfortable for them but, quite frankly, I didn't care.

My family, well, they didn't seem to care about my attitude, they were just pleased to have me home in one piece and alive. When I came back from Vietnam I still had 18 months of service obligation left; they knew I'd be going back to my duty station in California again in a month. What they didn't know was our battalion would return to Vietnam in six months. Perhaps if they'd known that, their casual attitude would have been erased, replaced by one of anxious concern.

Home

When I started as a freshman at St. Louis University I grew my hair long enough to fit in, and melded into the structure and fabric of the college lifestyle. Occasionally, I'd wear my old fatigue jacket to school, in anticipation of catching some lip from another student. I never did. Maybe it was because I looked as if I was always ready for a fight. Never once got asked any questions or received one welcome home greeting, not even a smart remark. I was essentially ignored.

I buried myself in school, studied hard, worked hard, and drank hard. I guess, in retrospect, I'd disconnected myself from my family and friends, and perhaps a better phrase would be a self-imposed isolationism. If anything, physically I interacted with people because I had to, but, mentally I kept them all at a distance. Before my time in Vietnam, I was very social and outgoing, and made friends easily. Afterward, my method for choosing friends became very selective. It's gotten better and, while I'm never going to be that happy-go-lucky guy again, as I was before Vietnam, I'm forcing myself back in that direction.

By the time Saigon fell in 1975, I was married with an 18-month-old daughter, owned a house, and had a full-time job. I was also in graduate school at that time. How I felt then is the same way I feel now—angry. Angry at the politicians who allowed us to engage in a conflict that was one of reaction as opposed to pro-action. The two-and-a-half million men and women who served in Vietnam ended up either coming back dead in an aluminum box; or injured, physically or mentally, or both. We sacrificed dearly just to watch those sacrifices erased by spineless politicians who were more concerned about their popularity and the number of votes they might lose. In doing so, by forcing the military to leave Vietnam, they discounted the humanity and the lives that were spent for naught. That topic raises my ire to an unhealthy level.

I guess I'm a classic case of Vietnam blowback. For years I drank too much and just about lost my marriage, but thank God my wife hung in there and didn't abandon me. I immersed myself in my job, which served to exacerbate the issues at home. Life was a real struggle for a long while. Then I went into the healthcare consulting business in the early 90s; it was both good and bad for me. In the late 90s Clinton began to make changes to the Medicare regulations and subsequently destroyed my business revenue stream. My partner and I also owned two home health agencies, and at Christmas time in 1998 I laid my entire healthcare staff off, closing the doors to our businesses permanently. My wife and I moved to Boston, as she had taken a job as a flight attendant for American Airlines. I started writing short stories, eventually graduated to writing novels. This fulfilled a life-long goal of mine. I took a job with American Airlines as well, which gave me minimal responsibility and allowed my mind to focus on my writing. On September 11, 2001, we were living in Boston when that brutal act of terrorism took place. This event was like a triggering mechanism in me. After that event I started having flashbacks along with intense bursts of anger and restless, sleepless nights. I hated to go to bed at night, for my dreams were peppered with the horrors of Vietnam . . . not good.

About five years ago I went to the VA in Boston and sought help. Those therapists with whom I dealt were excellent. Since I've moved back to the St. Louis metro region I have not sought counseling, but I intend to do so once I am situated.

My wartime experiences changed me in a myriad of ways that I still find overwhelming. My entire demeanor has been altered. I don't trust folks the way I did pre-Vietnam, and I still, even though I've been through years of PTSD counseling, seem to have an underlying edge to my temperament when I interact with people. As I reflect on my adult life I see so many things I did, which were, in retrospect, missteps, yet they seemed okay back then. Now I understand that they were not psychologically healthy and at times physically harmful to me. For over 40 years I refused to stay in contact with my battalion buddies or show any active loyalty to my military unit. I never talked about my time in Vietnam with anyone; very few people knew I was even there. That was done with intent; I didn't want to engage the topic. These days I feel a bit more comfortable about it. I'm now willing to attend reunions to visit with my veteran friends. Now I'm willing to talk on the phone or visit on Facebook with them, but aside from that, I'd rather be by myself.

I have been to the Wall in Washington, DC, and saw the portable Wall in St. Peters, Missouri, back in the late 90s. All I can say regarding my experiences with the Wall is I lost several high school friends and football teammates, along with Battalion brothers in Vietnam. It grounded me when I touched their names chiseled in the granite. To me, their lives were stolen from them for a cause that ended up being no more than a political and economic football game.

My experience in Vietnam was that of a reactive. Our primary objective as Seabees was to build. We built camps, hospitals, airstrips, bridges, roads, etc. We traveled with and supported the Marines everywhere they went. If our camp was attacked, we fought, and then when things quieted down we went back to building—defensive, not offensive. This strategy was frustrating. I can't tell you how many times I wanted to go out on patrol with a Marine unit just to feel some marginal sense of closure. In a way, I felt guilty inside. I was trained to fight, but that training was only put into play when we were attacked. As I reflect,

I wish I could've done more to further our efforts to squash North Vietnam.

Looking back, the war during my lifetime, in my opinion, has been glamorized by Hollywood. In reality, it's a bloody, life-altering event for those of us who have to engage in it. If necessary, war should be fought, but only as the last resort, and only when all other attempts at peace fail. The Vietnam War, as you well know, was not a war in that day's conventional sense. Except for the intermittent bombings of Hanoi, we basically chased the VC and the NVA around South Vietnam. It was like chasing ants at a picnic, only with deadly consequences. There were no major boots on the ground offensive assaults ordered by our government on North Vietnam soil.

We were spit at, called baby killers, identified as scum for serving our country. It happened to me and to others as we came home from Vietnam. This was another huge emotional millstone hung around the neck of each veteran.

Behaving this way was popular and trendy, and the thing to do. To vilify the veteran for the perceived transgressions of our nation, to make them the sacrificial lamb, it was in vogue. This lasted for a while and then faded when we abandoned the people of South Vietnam.

The collateral emotional damage was huge and a nagging burden to those of us who suffered the experience of that war.

It was over in 1975 for those who taunted us: we had given in to the Jane Fonda academic mentality and the world began to pass us by. I know I tried to keep up, but it wasn't easy. School, raising a family, a job, success, it all became a part of the chemistry of who I needed to be.

Deep inside, in a place no one else gains access, lies an ember of unabated anger, mistrust, and uncertainty. It calls to me at times and now that I'm older with more time to reflect I can hear it more clearly.

As I think back, I remember, after I left Vietnam, focusing on the societal injustices doled out every day and in countless ways. I aimed my anger and hostilities towards it. I know now that my anger had deeper roots. The culture of politics, that of the anti-war movement, the fuzziness of our purpose to be in Vietnam, the absurd rules which applied as we fought the enemy, the carnage—all those things factored into my being.

I longed for the life I knew before I went off to war, but it was only a memory now and I struggled with that, still do at times. It stole my innocence. It made me face things I never knew I might have to face. It challenged my beliefs and my faith. It confused me to my core.

There is no happy ending. It will always be the way I see it. I know this to be true. I find bits of solace in what I do these days and tighten my bonds with others of like kind.

In general, life has been good to me and I will accept what it offers, but I will always wonder if complete peace will ever come to my spirit.

RD3 Marty Severson
Ops/Intel, U.S. Navy 1968-1972

I don't know Murray Kompton, never met the man. Yet for the last 31 years I have carried the following newspaper clipping in my wallet:

"The tenth anniversary of the fall of Saigon has suffused the air with a good deal of nonsense, most of it emanating from persons who never wore a uniform there.

We are being told that the war's veterans are at last getting the respect denied them until now. Might we not be excused for wondering how much they really care? Anyone who withholds and then confers honor upon a soldier as though it were some grace within his own keeping has a very small claim to an opinion worthy of consideration."
Murray Kompton - 1985
Vietnam Veteran

What Mr. Kompton says is true. I, as well as MANY other veterans, feel this way: always have, always will. The words "Welcome home," spoken or written, ring hollow. They only really mean something when they're spoken by an active military member or a veteran, especially a Vietnam veteran. When a Vietnam veteran comes up to you and says, "Welcome home, brother," you know you've just found another lifelong friend.

Three years ago, I was invited to a summer wedding in northern Wisconsin. It was held in a beautiful park with waterfalls and lots of foliage. The 100 or so people in attendance (of whom I knew maybe three or four) were, to say the least, earthy . . . lots of sandals, flowers, and long hair. To each his own. It was a nice affair.

As I sat down to eat at a round table that seats ten, a slender gentleman about my age (64), with spectacles, white hair, and a long ponytail asked if he could join us. "Of course!" I said, "By all means!" He was very polite and well-spoken. As we chatted he mentioned he was from San Francisco. I told him that I had been stationed in the Bay Area for a short time and traveled through there going home. "You were in the military, I take it?" he said. "Yes, I was," I replied. He then told me, "Well, you probably don't want to talk to me then. I was a draft dodger and anti-war protester." I looked down at my plate and thought to myself, "Marty, that was over 40 years ago. Let it go." I shrugged. And then the draft-dodging anti-war protester said to me, "Well, that was a long time ago, so for what it's worth, thank you for your service, and welcome home."

My mind was racing now. "For what it's worth."? Are you fucking serious? "FOR WHAT IT'S WORTH"! I stared at him for a long moment. I thought to myself, "Okay, asswipe, when your country called on you, you ran, yet here you are, over 40 years later, successful and enjoying all the freedoms this country has to offer. I can think of over 58,000 reasons to beat the living crap out of you right now, and every one of those reasons has a name on a wall in Washington. Just because you decided to dodge the draft doesn't mean the Army would take one less person. No. They took someone else. It is very possible that the name of someone else is on that wall instead of yours. So, when you say, 'Thank you for your service,' you goddamn well better mean it!

"And then you said, 'Welcome home.' Fuck you. It's very possible that you were one of the assholes at San Francisco International Airport screaming names and spitting at those of us just coming back from Vietnam. We were 19 or 20. All we wanted was to get home and let our families know we were okay. And now, over 40 years later, you say, 'Welcome home.'"

I immediately thought of what Murray Kompton said: "Anyone who withholds and then confers honor upon a soldier as though it were some grace within his own keeping has a very small claim to an opinion worthy of consideration." Then I thought to myself, "Marty, this man has a very small claim to an opinion worthy of consideration." So, I took a deep breath and slowly leaned forward, stared in his eyes, and quietly said, "Please pass the salt and pepper."

PO3 Stephen P. Green
Mobile Construction Battalion 3 (MCB-3)

I was a high school dropout. I quit high school halfway through my sophomore year. Within a year after that, I realized that I was headed nowhere. I turned 17 on a Sunday and enlisted in the Navy on Monday morning. Mom had to sign for me. I didn't know much about 'Nam at that time. Mom would only sign for me if it were the Navy; she didn't want me going to Vietnam. Little did we know of the Seabees. Boy, was she mad.

I completed Steelworker "A" school in 1966 and worked that rate my first two-and-a-half trips to Vietnam. The rest of the time I spent in Security Company, guarding the base at night.

I spent four years with the USN Seabees. I made four trips to Vietnam, with about 28 months in-country, starting in 1965 going through 1968.

My brother Mike also served in Vietnam with the Marine Corps in 1968 on a 105mm gun crew. Our mom was lucky: we both made it home without being wounded. My brother spoke about Vietnam all the time. I didn't for over 40 years, and this worried my family. Mike enjoyed telling stories to his nephews.

I got my orders to Vietnam at the age of seventeen-and-a-half. I joined MCB-3 in August 1965, and stayed with them for my entire enlistment.

The Navy Seabees took a young punk kid and made him the man he is today. It didn't take long in Vietnam to realize how much I had taken for granted back home, not to mention life and death. War is not pretty!

I was at Hue/Phu Bai for Tet '68. It all started on the morning of January 31. We came under rocket attack. One of the first rockets killed Larry Stangel and wounded several others. Larry was to have flown home that very day. We were being relieved by MCB-8, and some of our flights had already departed when the rockets hit.

Richard Blevins was killed the next day during a noon rocket attack, with others wounded. Richard was only hours away from catching his flight home. We had just gotten the "all clear" following a mortar/rocket attack. Blevins came up to me and said: "Hey, Green, I think I am going to get killed tonight." Without forethought, I told him "Blevins, if that is the way you feel, you go find yourself a hole by yourself and stay there tonight." The next morning Blevins came running up to me and said: "Hey, Green, I made it, and I am going home today." I said, "Good for you, Blevins. See, there was nothing to worry about." Noon that day, several more rockets and mortars impacted on the base. Richard Blevins was killed instantly, hours before he was to catch the "big bird" back to "The World."

My fourth and final trip home was the first time I had traveled back to the States without my battalion. I felt guilty leaving my brother Seabees behind. I had to get out of Vietnam because the Office of Naval Intelligence was busting about four fellows involved in the black market, accused of mailing pot to the states. They were trying to implicate me in this also. With the advice of my legal officer, I was cut orders for stateside duty. I had long fullfilled my obligation for Vietnam service, and my brother had recently arrived in-country. I later found out those fellows all got 30 years in federal prison. I spent my last few months in public works on San Nicolas Island.

I flew into LAX instead of Point Magu Naval Air Station as before on my three other trips returning stateside. Somehow the flight home was uneventful. I think I slept most of the time, but that changed at the airport. At LAX, I was spit at and called a baby killer. These were not rumors, because it happened to me. Good thing my soon-to-be first wife was there to pick me up or I would have ended up in jail. This took place in late 1968, and was my Welcome Home!

"Numb" best describes how I felt getting back to the States. I think I had already started to stuff the bad memories of Vietnam deep down inside of me where they would remain for over 40 years and couldn't hurt. Survivor's guilt hit every time the evening news showed the war on TV. I would have reenlisted, as I wanted revenge for my fallen brothers, but the soon-to-be first wife put a stop to that.

I could deal with the public scorn because after a short time you didn't wear your uniform off base, and you didn't tell anyone you were a Vietnam veteran. The biggest slap in the face came from WW II veterans at the VFW. They didn't want us to join their post. They considered us drugged up murderers and the first veterans to have lost a war. I have never gone back to the VFW or American Legion. Suddenly, after 9/11, Vietnam veterans became cool! Folks started saying, "Thank you for your service." Probably the same ones that had spit at us years earlier but were now feeling guilty. That comment rarely means anything to me unless I know the person. What touches my heart is when another veteran sticks out his hand or hugs me while saying, "Welcome home, brother!" I had no home to go to so I just moved on with life. I self-medicated with booze and drugs until about 2010, when I finally quit denying I needed help. After Vietnam I have not had a best friend, just can't make that kind of connection anymore.

Home

While on my second deployment to Vietnam at Chu Lai in 1966, my mom had a nervous breakdown. When I returned to the States in late 1966, Mom was in a state asylum. My older sister was now taking care of my younger brother and sister plus her own kids. Mom was only 43 years old when this happened, and she spent the remaining 34 years of her life in nursing homes. Due to this, I no longer had a home town to come home to. I tried visiting a couple of times, but it just wasn't the same. Folks just didn't seem to want to be around a Vietnam veteran. So, from that time until I left active duty, I stayed on the base or lived in the town by the base. I didn't have to return to Vietnam for my third and fourth deployments, but just didn't feel like I fit in stateside. I felt it was better to return to Vietnam with the brothers I knew.

I married my first wife after I left active duty, a marriage that ended after five years and should have never been in the first place. I was living with my second wife on the day Saigon fell. She walked in and asked why I was crying. I had to tell her, "All those lives lost and for nothing." What a shame. The military didn't lose the Vietnam War; our government did. The North was ready to throw in the towel when we left, a sadness that remains in every Vietnam veteran's heart.

Postscript

During our '67 - '68 deployment, we lost eight of our Seabee brothers. Our first losses were on the night of August 30, 1967. I was awakened by a loud explosion. My friend Ron Burling and I raced to the hut near us that had taken a direct hit. Jack Wilkinson and Doug Coker had been killed instantly. Doug Coker was a friend of Ron's. The screams of the wounded and the carnage of that night will be forever burned into my mind. Forever! The next day we found out that the attack had been by friendly fire. A mistake.

I hope whoever reads this takes into account the horror of what Ron and I saw while serving in Vietnam. Yes, we came home, but got no Welcome Home. Instead, we were shunned, spit at, and called baby killers. I lost contact with Ron 40-some years ago, and we have only recently reconnected. I was surprised to find Ron was not getting the help he earned and deserved from the VA. It is also my hope this helps to change that. We served together with HONOR!!!

I wrote this letter to Doug Coker. I hope he is somewhere where he can read it.

Yeoman Coker,

I bet in a thousand years you never expected to get a letter from me. Yes, it is the Seabee Steelworker, Petty Officer Green, the Seabee you had to work with laying airfield matting on the night crew in Phu Bai, Vietnam.

I know you hated being required to help us Seabees to get that job completed on time for the Marines unit to arrive. Must have been tough on a fleet sailor and yeoman to be suddenly working like a Seabee. I know you hated it, as you would remind us nightly of what a short-timer you were. And in return, I would tell you nightly that you were not going to make it out of Vietnam because Charlie had a bullet with your name on it.

On that fateful night, I was asleep only a few yards from you when that 105 round hit. Within minutes, I ran to your hut to assist, only to find that you and Wilkinson had been killed instantly, with many others wounded. We may never have been friends, but I deeply regret having said the things I said to you. I have never and will never make light of death again. I am asking you at this time to find it in your heart to forgive me. You gave your all for God and country, and I will never forget you for as long as I live.

Your Vietnam Brother in Arms,
Stephen Paul Green

Chapter 5 - Coming Home

We came home from Vietnam by commercial aircraft, military transports, and ship. We came alone. We were with other veterans but for the most part they were not known to us. We all shared a common experience and were brothers and strangers at the same time.

Some aspects of our return were universal to all returnees, but every soldier faced a different reception by the public, family, friends, girlfriends, and wives. Some returned to a wonderful reception and absolutely no trouble from the public at all. But, for some the reception was horrible: girlfriends with another man, friends who abandoned them, and a public that scorned them. Many were ignored. It was like the war never happened. Their time in Vietnam was a lost year. The world went on in their absence. The world didn't seem to care.

The second time I came home was very different from the first. It was the end of May 1972. I was in a bunker on perimeter guard the night before I left country. I flew to the States on an Air Force C-141 cargo plane. It took me three days to get to Chicago. First stop was Okinawa, where we had a long layover, then we were supposed to fly to Travis Air Force Base direct. I sat on a crate playing solitaire for most of the flight. Unfortunately, we blew an engine somewhere over the Pacific. We heard a loud bang and the plane rocked and rolled. I jumped to the window and saw the in-board starboard engine on fire. The crew was running around and jabbering excitedly on the intercom. I remember thinking, "I am going to die like this over the Pacific after all that time in Vietnam?" Well, the crew managed to put the fire out and we diverted to Hawaii for a change of planes. Another long delay.

When I finally arrived at San Francisco airport I had been traveling for three days. I hadn't changed my clothes (khakis with big sweat stains under the arms), shaved, washed, slept, or brushed my teeth since leaving the Orient three days earlier. At San Francisco, the airlines requested that we didn't wear a uniform on the flight: too many problems with protesters. When I first went to Vietnam you couldn't get the military standby rate unless you were in uniform. Times had changed in America.

I planned on cleaning up and putting on civilian clothes before looking for a flight home. But, as fate would have it, I decided to check the airlines first for available flights. It just so happened there was a flight leaving for Chicago any minute and it had one empty seat. I thought, "Oh, jeez, I stink and look like hell," but I wanted to get home, so I took it. Unfortunately, I wasn't able to call home and tell anyone I was in the States.

They had to roll the staircase back up to the plane and I boarded. Well, the seat was in first class, next to a starched Army major. He gave me a real dirty look as I took the empty seat next to him. I know I didn't make a good impression. He kept giving me dirty looks, so after I sat down I leaned over and looked at his ribbons. He was only wearing a National Defense ribbon and an Army Reserve ribbon. I leaned back in my chair, shook my head and ordered a drink. I had quite a few drinks on that flight, actually.

When I arrived at O'Hare I called home. No one answered. I called my brother: nothing. None of my friends was home either. So, I hailed a cab and went to my dad's house on the northwest side of Chicago. There was one small problem. The house was locked and I didn't have a key.

Well, I dragged a garbage can into the gangway, climbed up, and jimmied a bedroom window. Once inside, I went to the fridge for a beer. There wasn't any. Not much of a welcome home

Army Vietnam Vet: Coming back home was almost as bad as being there. It's hard to come home from war to a country that doesn't want you.

Army Vietnam Vet: I came home twice and neither homecoming was pleasant. I have shoved the memories so far back that I can no longer remember the details. Hope it stays that way.

Army Vietnam Vet: I was fortunate on my two returns; I did not encounter any grief. Maybe being an officer helped. I did go through New York on one trip home. Port Authority had a questionnaire that asked for occupation; I put "Trained Killer." I am probably still in some database somewhere

.

Army Vietnam Vet: When I left Oakland Army Overseas Returnee Processing Center in June 1970, I went straight to San Francisco International, in uniform, with several others. As we walked through the air terminal not a person said or did anything to us. We parted company and went our own separate ways to purchase tickets for our home destinations. We had been warned to possibly expect confrontations from anti-war protesters, but it never happened. And this was in the capital of anti-war protest!

Navy Vietnam Vet: I came home in November 1967. I wish I knew what was coming. I think I would have gone AWOL and stayed in 'Nam. I think the VC were friendlier.

Air Force Vietnam Vet: When I returned home from Vietnam in 1970 I was spit on and called all kinds of bad names which I won't repeat. To this day I still have that ill feeling in my stomach. It is something I just can't shake. I am thankful for those who come home these days. It is better for them.

Air Force Vietnam Vet: Met a Hawaiian Airlines stewardess at Sea-Tac. She was flying back to New York for a wedding. She "thanked me for my service" all the way to O'Hare in Chicago. The pilot of the big DC-9 was a 'Nam vet and gave us (all nine vets on board) unlimited liquor. It was a great welcome home!

Army Vietnam Vet: I was checked by customs. Very demeaning but not unexpected based on how our country treated its troops. The plane home stopped in Japan and at a gift shop in the airport I bought two mother and child woodcut pictures to give to my sister. Customs ripped through the wrapping on those and ransacked our bags. WTF, after all we did for them!

Army Vietnam Vet: It was the same thing down here in the South: name-calling, being hit with stuff. Couldn't go anywhere without the cops harassing you. No one but family would speak or talk to you. The scariest part of being in Vietnam was getting off that airplane in Knoxville, Tennessee.

Civilian: The homecoming you all received has bothered me all those years also. I was nine years old when we picked my uncle up at the airport in 1969, coming home from Vietnam. To be a child and witness your loved one being hit with tomatoes, spit on, and called names had a profound effect on me. I became ashamed of my country. It is long overdue to be thanked and appreciated for your service and I am truly grateful for all of you. God bless.

Army Vietnam Vet: My homecoming sucked. I mustered out after 21 months in the service. The car I purchased in Vietnam was not where it should have been. It would take another two weeks to get there. The only flight I could get was to Dallas, Texas. I had a ten-hour layover before I finally got to Chicago. There, a flower child made a terrible decision to spit on me. As he laid on the floor bleeding all over the place, a security guard suggested I get out of there, the police were on their way. It was January 1970, and I was rather tan. I boarded a bus to Rockford and a lady sitting beside me inquired about my tan. I told her I had just got home from Vietnam. She got up and moved. At the bus stop in Rockford, three guys started giving me a lot of grief. I thought I was going to get my ass whipped but I stood my ground. They settled for a bunch of lip service and finally went on their way. When I got home, I told my mother if she wanted a picture of me in this uniform, take it now. I changed clothes and took the uniform in the backyard and burned it.

Army Vietnam Vet: I remember my mom asking me, "What did they do with my son?" I thought she had lost it, until years later I asked myself the same question. Welcome home, Brothers and Sisters!

Navy Vietnam Vet: When I came home it was not the most pleasant experience I've had. People at the airport were not what I call friendly. But the most disappointing thing was when I got home there was no "Hello," no "Where have you been?". It was like I was never born. It was like I was a stranger in my hometown. Vietnam was my home and after two tours I wanted to go back. It was bad.

Vietnam Vet: I got home about midnight. Mom had some nice food all ready for me. I told her that was nice, but if she didn't mind I'd rather have a bologna and cheese sandwich and a glass of milk. She looked at me kind of weird, and I told her it'd been a year since I'd had either one. I think it was the simple, ordinary things that we missed the most.

Navy Vietnam Vet: I ate a whole head of iceberg lettuce, then three peaches, and a pint of Jack Daniel's bourbon! Then took a bath!

Australian Vietnam Vet: It happened to us Aussies as well. Our RSL (Returned Service League) wouldn't recognize us for years: "We weren't in a real war," they told us. Times change and now Vietnam veterans are in many executive positions in the RSL.

Army Vietnam Vet: I was taking a walk in my hometown when I met a guy I knew had been to 'Nam, he was the older brother of one of my younger brother's friends. We talked for a while about 'Nam and I told him only my parents came to the airport to pick me up. He said, "You were lucky, I had to find my own way home: opened the door and walked in."

Army Vietnam Vet: I felt so out of place around my own beloved family. They all shunned me, most wouldn't even speak to me. Some even asked me how many babies I had killed on the average day. They wanted to know why did I waste my time coming home, they didn't need a baby killer among the family. I was treated like a disgrace to the whole family, I was spat on like I was trash, so I just packed up my stuff and moved on, never bothering to look back. It was like a large open black space came into my life. But I survive each day as it goes by. God bless all our brothers who served in that hellhole!

Army Vietnam Vet: I got to Fort Lewis, Washington, on a Sunday morning at about 4:30 a.m. I called home, which was 6:30 a.m. central time. For a host of really dumb reasons, I didn't get to St. Louis until about 11:00 p.m. central time. After what we'd been through—and this is probably just an ego thing on my part—a brass band welcoming wouldn't have been good enough, but there to meet me was my mom and my now ex-wife. They were worn out from the long ordeal; we all were. But they were there and I'll never forget how wonderful it was to see them. PS, I'm still good friends with my ex.

Army Vietnam Vet: Only my mom and my wife met me at the Broome County airport. That was enough. My wife always talked about how she "pounced" on me. Dang, that woman loved me.

Army Vietnam Vet: When I came home, my dad (who had fought under Patton) gave me the good news; "Vietnam was not a real war," and commented I was not there for as long as he had been in Europe. I was sending home an allotment on E-5 salary and when I came home they had spent it all. Glad you are home. No welcome there nor at the VFW two miles away.

Navy Vietnam Vet: "Numb" is a good word, or "tired." Cab from the airport to home. Got a hug from Mom who made some calls telling people I was home. That's the last thing I would remember for several months. Everything seemed so surreal at home. My closest sibling told me, "Thank God you are home. You never should have been there."

Army Vietnam Vet: I was in the 199th Light Infantry Brigade. When I came home my father asked me what I did to keep from getting shot. I was taken aback by the question and felt uncomfortable. I thought a minute and said to him, "I was smart and I was lucky." A little confused, he asked me why I was smart. "Well, Dad, I was smart because I knew if I could find a tree as wide as me I would be safe." "Why were you lucky?" "Because I found enough of them." He never asked me about 'Nam again.

Army Vietnam Vet: I returned from Vietnam wearing Class As because I was using a military air ticket. When I landed at Boston's Logan Airport, I was sneered at and called a baby killer by some college kid types. I ignored them.

Army Vietnam Vet: When I arrived in San Francisco with many of my fellow 'Nam vets, I was on the phone calling home when a jerk grabbed my shoulder and shouted, "Baby killer!" and tried to spit on me. I spun around and instinctively hit him in the nose, breaking it, and was getting ready to hit him again when another vet shouted, "No!" That brought me back to where we were and I stopped before killing the little punk. He ran for the cops but I was quickly surrounded by all the other troops and he had no idea who had hit him. We stuck together even after we made it back to the United States. 'Nam vets will always be my brothers.

Army Vietnam Vet: As soon as I landed in the States I went into the airport cafe and tried to get something to eat. I was told, "We don't serve your kind in here," and when I went outside I got hit with a balloon filled with piss.

Army Vietnam Vet: I was medevaced home. I was carried from the plane to a hospital Quonset hut near the tarmac by military personnel. Was another two months before I went home on a convalescent leave. Still looking for the real world.

Army Vietnam Vet: Western Airlines was on strike. There were only two other airlines that serviced LAX: TWA and United. They were swamped with people. I bought a first-class ticket on TWA but spent over 12 hours waiting for the flight. I had arrived at Sea-Tac at 0700 and boarded the flight to LAX at 1835. I had to put up with the name-calling, spitting, spilled drinks, and crap tossed at me. Got home and the next day my dad had my uniform dry cleaned.

Army Vietnam Vet: We came home alone, not with the brothers we were with for so long, not with the brothers that arrived in-country after us. We came home alone. One day in conflict, next day at Cam Ranh Bay or another disembarkation location with brothers we did not know but understood. Then we boarded a plane and went home alone. One day we were in battle, the next we were at home, alone. We had our parents, siblings, and high school friends, but we were alone. Now they go as a unit, return as a unit, and are not alone. Ours was different. We were alone.

Marine Vietnam Vet: Came home on a very, very old ship from Da Nang Harbor. In the middle of the night, while most were asleep, we awakened to a loud "thud, thud, thud," which became a loud rumble.

Then one could feel movement. We were going home. There was no cheering. Quietest group of Marines I was ever with. When dawn broke all one could see was water, and it felt very cold . . . strange, surreal feeling . . . October 1969. There never was any cheering, not even when we docked in San Diego some three weeks later. Just quiet.

Army Vietnam Vet: I came home in December 1972, on a C-141 medevac flight with a whole planeload of shot-up, busted up boys coming out of 'Nam. We parked next to the chicken wire fence on the tarmac at Travis. When they let the back door down and they carried us out on our litters, before we got our first breath of air drawn into our lungs of being back in The World, protesters were screaming and calling us names from behind the fence. I will never understand why the base commander let that happen. It would never happen today. The guys carrying me said that they knew just where this flight was coming from and when it arrived, and they were always there to greet us. One guy said, "I guess they don't know that I was just fully armed just the other day."

Chapter 6 - Air Force Stories

There were ten major Air Force installations in South Vietnam. These bases housed combat aircraft such as F-4s, F-100s, A-7 Skyraiders, and AC-47 Spooky gunships. They were also the home of military transport, reconnaissance aircraft, and helicopters of various types. Major logistical centers, repair facilities, and hospitals also had a large presence on these bases.

The Air Force provided close air support for Army and Marine ground units throughout Vietnam, while also flying strategic bombing missions in South and North Vietnam. Their mission also included transporting troops and supplies, and spraying herbicides such as Agent Orange. Air Force helicopters made many emergency rescues of downed airmen.

Air Force aircraft transported the wounded from Vietnam to Japan, the Philippines, and the United States.

B-52 bombers flew countless missions over North and South Vietnam from bases in Guam, Okinawa, and Thailand. During the war, it is estimated that 80% of all Air Force strikes over North Vietnam were launched out of Thailand. At the peak of Air Force involvement in 1969 there were more Air Force personnel serving in Thailand supporting the war effort than in South Vietnam.

Throughout the war the Air Force flew over five million sorties over South Vietnam, North Vietnam, Laos, and Cambodia.[1] The Air Force lost 2,251 aircraft during the Vietnam War. Two thousand, five hundred and eighty-six Air Force personnel died during the war, in addition to 931 wounded requiring hospitalization, and 2,518 wounded not requiring hospitalization.[2]

1. USAF Operations Report, 30 November 1973
2. Casualty figures from Defense Casualty Analysis System (DCAS)

Staff Sergeant Jimmy Taylor
U.S. Air Force Combat Weather Team, Camp Evans

I had dropped out of school in 1967 and had taken a job, then I went back to school in late '67 but had the draft on me. After the first of the year in 1968, I had gone for my second physical. That was when I pretty much knew I would be getting drafted. In March 1968, I made the decision to join the Air Force. I decided that I was not going to be drafted one way or the other. I eliminated the Marines as a possibility mainly because I saw "The D.I." with Jack Webb when I was a kid, and I didn't want to go through that.

So, I joined the Air Force. I left for induction on April 14, 1968, which was Easter Sunday morning. I went to Jacksonville, spent the night, and then the next morning went through the process. I remember what a deal that was. We were in a room waiting to talk to a woman who was typing up our information, and we were signing documents. I remember that the waiting room was full of Life and Time magazines. It was April 1968, right after the Tet Offensive. Every magazine had a front-page cover of the war. And I thought to myself sitting there, "Why the hell don't they get rid of these magazines?"

I remember looking around the room and thinking, "Jeez, I wonder how many of these guys are not coming home. I wonder if that guy is going and maybe he's coming back and this guy . . ." Anyway, I sit down to be interviewed by this woman. She's typing forms and asking questions, and we're signing things. "Have you ever been a member of Ku Klux Klan? Have you ever been a member of a communist party?" I remember that form, particularly, as just a long list of organizations that had you been a member of, you probably wouldn't have gotten drafted. But the one thing I remember was the woman. I guess she was in her 20s, not much older than me, maybe in her 30s. But as she was asking these questions she stopped, folded her arms down on the carriage of the typewriter, and looked at me. And in a quiet voice, she said, "Now, who do you want to notify in case of your death?"

And I remember it took me aback. It made me kind of wake up to what I was doing. And, of course, I told her, but what I really remember is how her attitude and her voice changed, and the emotion in her voice when she asked that. Her whole being changed. When it was over, she put her arms on the typewriter like she wanted to hug me. It was an amazing moment, and it stuck with me all these years.

I coasted through basic training and had my choice of jobs. I could take anything on the board because of my test scores. I had three picks. My first choice was air traffic control because I could go to Biloxi, Mississippi. My other two choices were weather, and photo mapping. The Air Force decided I would be a weatherman, and I couldn't have been happier.

My training was at Chanute Air Force Base in Illinois, and after graduation I was given a choice of duty stations. I chose Tyndall Air Force Base in Florida, and I might have chosen Robins . . . I forget. But that's where they sent me. To Robins, 90 miles from home. But two years later, I got orders to Vietnam.

I was home at my parents' motel in Dawson, Georgia, when a call came in for me. It was my commander's executive officer, a guy named Jim Goldey. He said, "Jim? This is Jim Goldey. I got your orders. And I got to tell you, we got bad news, and really bad news." And I said, "Well, tell me the bad news." He said, "You're going to Vietnam." I said, "Damn, Jim. What's worse than that?" And he said, "Well, we can't pronounce it. And not only can we not pronounce it, but we also can't find it on a map." That was not good news, because I'd always said, "Hell, if I go to Vietnam, they're going to send me to Tan Son Nhut, Da Nang, or Cam Ranh Bay or somewhere." But they sent me to a place called Hue/Phu Bai and attached me to the 101st Airborne Division.

Well, I arrived in Los Angeles and proceeded to party all night, and by the time I caught a bus to Norton Air Force Base I had missed my flight.

I flew into Da Nang two days late. Nobody was there to meet me, so I slept on my duffel bag at Da Nang airfield. I finally got a hold of somebody and he came and got me. He took me to Camp Horn, a little Army field up above Da Nang. And that's where our group was headquartered with the Army. I spent a day or two there getting acclimated, getting some clothes and stuff. Then they took me back to Da Nang to catch a C-130 up to Hue. That's how I got to Phu Bai. I eventually wound up at Camp Evans where I spent most of my tour. I was moved to Khe Sanh during the Lam Son 719 offensive into Laos.

Coming Home

I had sent most of my clothes and personal stuff home before I left Vietnam. I went to the Da Nang PX and bought a pair of black lace shoes for the flight home, and they wore a blister on my foot. I took a shower the morning before I left without shower shoes. And that blister, 19 or 20 hours later, when I got back to the United States, had become infected. And it was up both my legs, to my waist. It was the greenest, ugliest looking fungus you have ever seen. That's how nasty it was in Vietnam. The air was filthy, everything was.

I left on a Pan Am charter flight. I remember early that morning seeing one of my buddies at the weather station talking to one of the stewardesses on that plane. He called me over and introduced me to her. We got on the plane, and I remember it was a somber thing. There was no laughter or talking.

We took off and minutes go by and there's no noise: quiet, just quiet as a church. And then you heard the ding of the smoking lamp and then the pilot comes on the intercom and he says, "Gentlemen, we have just left Vietnam airspace. And we're going home." When he said that, it just erupted. Three hundred and fifty guys as one just started yelling and clapping and cheering. After everyone settled down, the stewardess came on the intercom and said, "Jimmy Taylor." And my ears perked up. And the stewardess was saying, "Have we got an Air Force sergeant on board named Jimmy Taylor? Sergeant Taylor, where are you? Would you stand up, please?" And I thought, "What in the hell is this?" And so I stood up, and I moved out into the aisle. And this beautiful stewardess came down the aisle with a great big smile on her face and all the guys looking, and when she got to me, she didn't say a word. She just took me in her arms and just kind of bent me over and gave me one of these huge kisses, a big wet kiss. Just the sweetest thing you ever saw. She was a knockout. And she said, "That's from all of the guys at the weather station in Da Nang." And I didn't know one of them. Never knew any of them. I just stayed with them that two or three days while I was waiting to fly out.

We flew into Norton where we went through customs. Then they bused us over to LAX, and that's where they turned us loose. I remember it was midnight, and there was nobody there. No demonstrations, nothing. Nobody even looked at us. But again, nobody walked up and said, "Welcome home," either. Nothing. No, "Did you just come back from Vietnam? Hey, welcome home. Let me buy you a drink." Nothing.

I had a 7:30 a.m. flight, so another guy who was on the plane with me decided to split a room rather than wait at the airport all night. I called a guy I knew who had been a helicopter pilot with the 101st, and he came over, and we drank and talked all night. His name was John Michaelson. A wonderful guy.

The next day we went back to LAX, I got me a Bloody Mary, and finally got on the plane about 7:30 a.m. for take-off. The stewardess came back to where I was sitting and put a bottle of champagne down in the seat next to me, between a guy who also was from Albany, Georgia, and me.

By the time I landed in Albany, after about 48 hours and a lot of alcohol, I was out like a light. Everybody got off the plane. My daddy said to my best friend, "Where the hell do you reckon he is?" And Rusty said, "Well, he's probably in Atlanta having fun." Well, Rusty knew me pretty well. About that time, a guy walked up, and said, "You all waiting on an Air Force sergeant?" My daddy said, "Yeah." He said, "Well, you better get a stretcher. He's up in the plane." So they carried me off the plane. And there were all my friends and my family and everything, a great welcome home.

Home

I have a lot of empathy over the years for people that did not have a good welcome. I never could see that. I never could figure that out. I never could understand why people were on a soldier's ass. Did they want us all to run away? Did they want us all to betray our country? Did they want us all to desert, to basically commit treason? Did they want us to be traitors? I never did understand it. We were there because they told us to go there.

A lot of guys said, "I want to go to Vietnam," and volunteered to go to Vietnam. I thought that was a great thing. Personally, I volunteered to do things when I got to Vietnam, like go to Khe Sanh, which put me in harm's way. But I never had any inkling of volunteering to go to Vietnam before I went there.

But I also think that the vast majority did not volunteer to go to Vietnam. They were in Vietnam because they were ordered to go. The problem lay with the people that sent us there. We were not invaded by Vietnam. It seemed to me that we were there trying to put our stamp on a country for vague reasons. And I look back at pictures I have now of peasants plowing a damn rice paddy. And my thought was then as it is now: those people didn't give a rat's ass who their president was. All they knew is, for 500 years their country had been torn apart by war. And all they wanted was to be left alone and grow a patch of rice.

And we went over there. I wish I could figure it out. I have read every book I could about it. I've been a student of history all my life. I'm not a smart guy, I'm no intellectual, but one thing I learned about reading American history and world history is this: unless you piss off the American people, they're isolationists. By nature. We have never fought a war to a successful outcome in which we were not attacked. And Ho Chi Minh, who was educated in the west, probably knew as much about the United States and the history of the United States as General Westmorland, McNamara, or anybody else. He knew that as long as he didn't do something to really piss the Americans off, sooner or later he'd prevail. And he would prevail because he had unlimited people. And he would keep pouring people in there just as long as it took. If it took 15 years, that's how long he would take. They were just going to keep doing it.

I look on Facebook, and I noticed that a lot of guys on those sites say, "Finally, I'm able to talk about going to Vietnam." I've found that to be true with me. I never mentioned it much. I didn't talk about it much. I didn't want to bring it up in front of a guy who laid in a hole for a year getting shot at. But then again, nobody ever asked. And I don't ever remember being thanked for my service until two or three years ago. After all those years.

And it's kind of sad, kind of sad. And I feel for guys that say, "I don't want to live around people. I live out in the country." I noticed a guy who said that the other night. And a lot of them live alone, you know, away from people. It's amazing. A lot of them feel put upon, and I kind of understand that. I abhor these things. I get a lot of . . . you know, around my buddies, they say, "You didn't do anything." We went . . . we did what we were asked. Like a patriot. I get teary-eyed sometimes thinking about it.

I wish it could have been better for those of us that served. It was a terrible thing. This war was about us. It really was about the men and women that went over there.

I basically tell people I had a great experience. And I did. But I look at pictures of me, and there's some that you can see I was not the happiest camper, maybe a little fear in my eyes or whatever. But for all my life, since I went to Vietnam, I'd tell a story, a funny story about something that happened to me while I was over there. But I didn't go into a whole lot of detail. And someone would say, "I'm tired of you talking about it," or some guy would say, "You didn't do anything. I was in the bush." Yada, yada, yada.

I've avoided those circumstances. But you know, they shot at me the whole time I was over there, in some fashion, either a mortar or rocket. They fired at me when I got there, and they fired until I left. It didn't happen constantly, but it happened regularly. How many people have been through that, you know? Do you know what I'm saying? But I respect those guys that went over there and were in the bush. I joined the Air Force because I didn't want to do that.

There were no Audie Murphys that they made movies about. There were Medal of Honor winners, and there were heroes, and there were a lot of brave guys. But there were no bond rallies or tours. They didn't bring us back as heroes like they did the Iwo Jima Marines. We got none of that. And I know none of them asked for any glory, but at the same time they didn't ask to be treated with disrespect. Regardless of any spitting incidents and cursing and that sort of thing, there were no parades. And now, they're beginning to have them. And it's good for these guys, you know? It's good. I mean, that's the way you . . . that's something that happened all during our history. They welcome the soldier back. And boy, that was a big-ass deal. It was big from the Revolution on.

And it did not happen in Korea, and it did not happen in Vietnam. I think we have a kinship with the Korean veterans because they got the same kind of treatment. It's been hard to overcome that, especially as you get older. This thing becomes a lot more important to us today when we're 65, 68, 69, 70 years old than when we were 30. I mean, when we were 30, hell, we're trying to raise a family and make a living or whatever. But now, as we get older, it becomes a little more important. You have a lot of memories bottled up, and you want to tell the story.

I was lucky. I was lucky. I was lucky I didn't have the kind of homecoming a lot of the guys had. But once I got back home, it went away.

Almost this entire experience went away for 30 years. And only my very closest friends ever asked me what it was like. Nobody ever said, "What did you do?" They would listen to my stories if I wanted to tell a story. But nobody ever asked about it. My parents didn't ask. They didn't want to know. They didn't want to hear it. My mama stayed an alcoholic for ten years. She became an alcoholic when I left, and she stayed an alcoholic for ten years. And that was it.

Like all other things connected with the Vietnam War, we went by ourselves. We never had to get on a troopship. I mean, a lot of them did. But it was a war of solitude; I always thought that. There was a kind of aloneness to the war that, as you look back, this was not the case in World War II. They shipped out as units. I would say 99% of the time we went as an individual.

And I never told anybody. But I've been to the VA, I've been to the mental health clinic. But I never could make myself say to them, "I think I'm suffering from posttraumatic stress syndrome." Because I wasn't shot at in a fox hole. I didn't wake up screaming with demons. I dream about it now more than ever. I dream about things that went on now that never bothered me. But I wouldn't trade my year in Vietnam for anything.

Master Sergeant Glenn Murdock
56th Air Commando Wing
76th Military Airlift Squadron

I joined the Air Force in August 1966. The reason I did that is I had my draft notice to go into the Army. I was supposed to report for my draft physical but I had already completed it for the Air Force. I attended basic training at Lackland Air Force Base and on my third day I got my notification for induction in the Army! After basic I was trained as a security policeman. It is not a job I picked but was assigned to. After training, my first duty station was at Eglin Air Force Base in Florida.

While at Eglin in September 1967, I received orders to Thailand. It didn't bother me. It's part of the job that you're committed to do. When I arrived in Thailand I was assigned to the 56th Air Commando Wing at Nakhon Phanom, which is on the Laotian border, about 15 miles from the Ho Chi Minh trail. My tour of duty there was for one year. I was in the base security force, and the Pathet Lao would come across the border and steal PSP (perforated steel planking) which we used for the runway, as we didn't have asphalt there. We were so close to the border and the Ho Chi Minh trail that at night I would sit in the bomb dump and write letters, because it was so bright from the flares they were dropping over the trail.

In September 1968, I came home from my tour in Thailand. I flew on World Airlines from Bangkok to Tan Son Nhut, picked up more troops, and then I think we went to Hickam Field in Hawaii, and finally San Francisco. We didn't get off the plane during the entire time from Bangkok to San Francisco! Needless to say, no one on that flight smelled very good! When we arrived at San Francisco airport we were met by protesters. It was a shock. I never expected that. We were all in uniform, so to avoid problems from the protesters they took us to a private exit, with a police escort. I just gave up a year of my life to put up with this? I mean, what the hell were we there for? We didn't have the support that they have today. We were either treated poorly or ignored like nothing happened.

After my tour in Thailand I was assigned to security at a B-52 base in Plattsburg, New York. In 1970 my enlistment was about up and they wanted me to re-enlist and I said, "I will reenlist if you cross-train me."

And so, I was sent to Shepard Air Force Base and trained as a loadmaster. After that I went direct to my duty assignment with the 76th Military Airlift Squadron at Charleston Air Force Base for on-the-job training. I was upgraded to air crew when I was in Charleston, and from 1970 to 1979 I flew on C-141 aircraft as a loadmaster. I spent the first five years of my loadmaster career flying back and forth to Vietnam.

We were based in Charleston and we would fly to Elmendorf, Alaska,

and another crew would take the plane to Yokota. We would stay at Elmendorf for a day or two and either crew a plane flying in from Japan back to the east coast or we would crew a plane heading west to Japan and Vietnam. We would fly supplies into Vietnam, and wounded to Yokota. We just kept the flow of planes going. These flights were considered TDY (Temporary Duty).

We flew into Da Nang, Tan Son Nhut, and Cam Ranh Bay. I remember flying into Cam Ranh and we would do what we call a penetration approach. We would be flying at 40,000 - 45,000 feet and we would do a screaming ass descent! That way we wouldn't get any ground fire. I remember going into Cam Ranh one day, and I hadn't been flying but a year or two, and my buddy was a loadmaster with me. I was looking out the troop door. I said, "What the hell was that red stuff?" and he says, "That's ground fire." It was pretty unique.

We flew casualties and the dead home. From the front to the back of the plane there was nothing but flag-draped caskets, actually transfer cases. I stared at those transfer cases for hours and hours. That was the toughest part of my 22 years in the Air Force. On each transfer case there was an access panel with a death certificate in it. I read one once and never read another.

I still think about the evac. missions with wounded guys. I remember once, I was on air evac. mission out of Tan Son Nhut and we had this one guy, Army guy. His whole intestines were lying on his chest with wet towels. I will never forget that. And they had to keep it from drying out. Our airplane just sucked up the moisture so they had to keep putting the wet towels on him to keep the intestines damp so they didn't dry out. I'll never forget that. Those were pretty rough flights. You just see all these guys who were badly wounded. Not the guys that got shot in the leg: these guys were all messed up. We had nurses on our airplane but I can't remember how many. We carried a bunch of litters, something like 54 on our aircraft.

Sometimes, we would get what they call "blue bart," which was somebody escorting a body. It could have been a buddy, his brother, or could have been his dad. And that was the hardest part, I tell you, that was the worst. We had jump seats for the escort to sit on and we treated them like a king.

Those flights were long and lonely. We would fly them to Delaware because there was a mortuary there.

Oh, man, I guess it's just hard to explain how I felt back then. Today, I'm a member of the Patriot Guard and I feel like I'm giving back to what was taken away from me. It's kind of hard to explain. I pretty much put it behind me, but when I see a casket with a flag on it at the Patriot Guard, it brings back those memories.

Coming Home

I went back and forth from the United States to Vietnam and Thailand constantly for several years. The people here had no idea what was going on over there. Nobody ever asked me about what I did. They just said you were in Air Force, probably didn't do much, you weren't involved in anything. You just flew around in planes. They didn't know what the hell was going on. No one did unless they were there. We didn't talk about it. I never did do a whole a lot of talking about the war until much later. You can't change yesterday. You can only change today or tomorrow.

Home

In 1975, when Vietnam was falling, we were so busy. We flew out the refugees and we didn't use seats. The capacity of our planes was like 154 paratroopers, and here we are with, like, 400 refugees stuffed in the back of the airplane. We would run our straps across the floor and they would have to hold on to them when we took off. I mean, what are you going to do? You have to do what you have to do. It was against every damn regulation in the Air Force but we did it.

We were picking them up at Tan Son Nhut for the most part. Every once in a while we were sent to Bien Hoa or Cam Ranh. But most of our activities were out of Tan Son Nhut. When Saigon fell, I think they had every single C-141 in the world over there. All I could think was, "What a waste."

In Thailand, they defoliated our whole perimeter with Agent Orange and I was a perimeter guard, so I was in the middle of it. Now I have glaucoma and an enlarged prostate, but the government won't do anything about it.

I have been to a few unit reunions, but most of the guys were not from the Vietnam War. These were the next generation. It was all right but I'd rather stay home. I still keep in contact with some of the guys from Thailand and Vietnam through Facebook.

I have never been to the Wall in Washington but I have been to the Moving Wall. From what I understand it is just about the same as the one in DC. I would never go to the Wall in Washington. I've seen all those names of the guys that didn't come home and it's just kind of hard on you, you know. Fifty-eight thousand of them; that's a lot of people.

I personally believe we should never have pulled out of Vietnam. We should have kept going. My opinion is that the war wasn't run by our general staff. It was run by politicians and they sure botched it up. But I would go back to visit Vietnam in a minute. And, if there was another war over there, I would go back, too, if they let me. I spent 22 years in the Air Force and I would go back.

I consumed quite a lot of alcohol, which is something I carried back with me from 'Nam.

Sergeant James P. O'Connor

35th Combat Support Group, U.S. Air Force

I have two older brothers that served in the Air Force in the 1960s. My brother Mike and I joined the Air Force under the buddy system in August 1966. After basic training Mike was stationed in Minot, ND, as a missile control wiring specialist. Our brother Ronnie had been in the Air Force since 1956. His specialty was air traffic control. Ron was ten years my senior. Neither one of my brothers served in Vietnam. Ronnie,

however, was in Udorn, Thailand, when I was in Vietnam, 1968-69. Ronnie was one of the guys working in the control tower that kept the aircraft safe during takeoff and landing. Many sorties were launched out of Thailand to bomb North Vietnam, Laos, and Cambodia. Ron and I kept in touch through letters. Mike never spoke about 'Nam because he never left the US. Ronnie and I never talked about Vietnam.

I served in the Air Force from August 29, 1966, to August 28, 1970. My brother Mike had received a draft notice for the Army, so he enlisted in the Air Force and I enlisted with him. They took him right away and me three weeks later.

My assignments included March Air Force Base in Riverside California, Guam in 1967, March AFB for another six months, Lackland AFB for two weeks of so-called combat training, then Vietnam with the 35th Combat Support Group in Phan Rang from April 1968. After Vietnam, my final duty assignment was at Andrews Air Force Base in

Maryland with the 1002nd Security Police Squadron. My Air Force specialty code was Security Police.

When I received my orders for Vietnam I had mixed feelings. Part of me wanted to get over there and part of me was scared. I was dealing with a whirlwind of different thoughts and emotions. I even pictured myself being shot at by gooks and returning fire. But that didn't seem real at that time.

Coming Home

Coming home, my departure from Phan Rang was somewhat difficult. I was happy to be going back to The World but felt like I abandoned the guys I left behind. While waiting for our flight home, an announcement came over the loudspeakers that Phan Rang was under attack. Ground infiltration and heavy rocket fire. One K-9 trooper was killed. I was filled with anxiety.

On the flight back to the States, I kept thinking about the attack on Phan Rang. I was so glad to leave but I had mixed feelings, like I wanted to stay. I was happy to see round-eyed girls again. The majority of the 23-hour flight was spent sleeping when I could, or at least attempting to sleep. When I departed Cam Ranh Bay, it was 120 degrees with 100% humidity; when I landed in Seattle, Washington, it was 85, and we were freezing. From there I caught a continuing flight to Washington National Airport.

My mother, father, and two younger siblings greeted me at the airport. No one said one word to me about Vietnam. The only thought I had at that time was rejection.

Being back in The World felt unreal. I was ignored more than anything. I caught looks from some people even though I had returned home in civilian clothing. Some recognized us by our short hair, others because of our darkened skin from the sun, which I referred to as Big Red. In my hometown, I just kept myself isolated. I always kept myself in the background. Never spoke much, just was always vigilant, and listened more than spoke. I was never able to contact any of my former friends. I had no idea where they were. Six months after returning from

Vietnam I met up with a fellow Vietnam vet, Bill Dowd. He and I met when in Vietnam when my tour was half finished. He and I were stationed at Andrews Air Force Base after Vietnam. We ended up partying for a few weekends then we were eventually discharged and went our separate ways. I never heard from him again.

Home

It was easy for me to participate in road rage incidents. I consumed quite a lot of alcohol, which is something I carried back with me from 'Nam. I worked the rest of my life and put so much of myself into it that 'Nam was blocked out. I was too busy working and raising a family.

There was a significant disconnect from family and friends. I kept myself involved with my jobs, giving them 110%. Vietnam did change me, but I didn't realize it until I was diagnosed with PTSD in 2005.

In 2003, I retired from 30+ years civil service at the Washington Navy Yard. It was then that I had plenty of time on my hands to do nothing but drink all day and part of the night. All of a sudden, the sound of choppers, loud aircraft noise, and backfires brought flashbacks of 'Nam. A licensed clinical social worker named Angela Hoag diagnosed me with PTSD. I went through many hours of both individual and group therapy and am still in therapy for PTSD to this day.

In 1975, when Saigon fell I was working at the Washington Navy Yard as a plumber. I was repairing a 16" water main when I heard the news. I experienced profound disappointment when I heard that the NVA had won. My first thought was, "What the fuck happened?" We had a chance to win that war and set South Vietnam free. Even decades after returning from 'Nam, I was still a very troubled person. Couldn't figure out why I went through my first divorce in 1985 and through a second divorce in 1998. I have been on my own since then.

I was completely burned out when I retired in 2003. I kept on drinking all the time, started having flashbacks: I was back in 'Nam. I had visions of the guard towers, the perimeter of Phan Rang. Dreamed of a chopper that blew up and of watching several people burn to death, and I couldn't do a damn thing to help them. Still have a constant vision of a dead gook body on the barbed wire fence with his guts blown out.

I remember searching in a swampy area for gooks, and being lifted up in the water and dropping back down. Never did know what that was. I was stung by a scorpion in the chest, just below the breastbone. Thought I was a dead man. It turned out the scorpion was a yellow one. The doctor at the base clinic said they were not deadly to humans, but I would be sick for about four days. He gave me an antivenom shot and told me to keep an ice pack on it for the next four days. Because of the excessive heat in 'Nam, I asked him, "How do you propose I keep ice on this when it's 98 degrees with high humidity?" He left and came back with a surgeon's glove with dry ice in it.

Now, adding further depression to the whole war, in 2010 I was diagnosed with prostate cancer. A biopsy revealed nine out of 12 samples were cancerous. The prostate was removed and a biopsy of the removed prostate showed signs of leakage. Thus, 39 consecutive radiation therapy sessions were performed. Through all this, I have had trouble with urine leakage, urgency issues. Now the inside of my bladder looks like strawberry fields. That is irritation burn marks from the radiation. Also, burn marks were found in the rectum from the radiation. All this because of exposure to Agent Orange. Just another way Vietnam won't leave me alone.

Vietnam was detrimental to my mental and physical health. Flashbacks and physical health issues are constant reminders that we were there.

In retrospect, I will thank you in advance for asking me these things. I am in tears at the moment, but have the mental thing under control. Sometimes, letting the water stream from your eyes is good therapy. Hoping for a happy day.

Chapter 7 - Adjustment

Home. Home. Finally, we were home! Most us came home individually. We stepped from one alien world into another: The World as we knew it had changed. Our friends and families had changed. The World we had looked forward to for the last year was now hostile or apathetic. It was depressing. It wasn't how we imagined it would be.

Slowly, we started to put our lives back together. Many were able to put Vietnam away in the back of their minds, buried in a place where it lay dormant for many years. Some were never able to put it away. Our adjustment was complicated, just like the war. No one wanted to hear about it. It was old news. We drifted into the shadows, and in the shadows we remained.

I was finally home from Vietnam for good but I didn't feel comfortable there. I was still in the Army and had six months left on my enlistment. I was sent to Fort Huachuca, Arizona, to finish out my time. I checked in at Fort Huachuca and found out that there were so many guys coming back from Vietnam that they didn't know what to do with us. So, we had morning formation at 0900 and were released for the day at 0930. In the desert with nothing to do. It was horrible. I hated every day there.

When my enlistment was up and I was discharged, I came home a lost young man. I didn't know who I was or where I was going. I always felt I was up in the corner of a room, at the ceiling, looking down at myself, watching everything I did. I felt very alone. I had just turned 21 and I tried to connect with old friends, but it just didn't work. I was angry. A friend from grade school and high school stopped by and I just couldn't talk with him. I was distant and, I guess, rude. He left and I didn't see him again for many years. Last year I reconnected with him and apologized for my behavior.

I had no girlfriend. I didn't know how to relate to women. I joined the Army two days after my eighteenth birthday, and was in Vietnam less than a year later. I was 20 when I finally came home. I didn't know how to date. While in the service my contacts with women were not of the most wholesome types, if you know what I mean.

My first job home was as an aide in the locked unit of a psychiatric hospital in Chicago's Uptown neighborhood. Not a good area back then. Basically, I was a bouncer there, fighting and putting people in restraints every day for nine months. The stress took hold and my Vietnam-related stomach issues reared their ugly head. I wound up in the hospital for two weeks, and quit that job to attend school full time.

I withdrew. There was a small corner store at the end of our block, Wagner's Grocery. I used to go there for cigarettes and groceries for my mother, and of course I had an occasional pop or candy for myself. Charlie and Helen Wagner were wonderful people. They loved the neighborhood kids and would give us free candy now and then. They watched us grow up. Charlie died while I was in Vietnam. When Helen heard I was back home she told everyone that she wanted to see me. I was one of "her kids." She knew me from when I was old enough to walk to the store with my mother. I never could bring myself to see her. I didn't want to see anyone. She died a couple years later. I never saw her after Vietnam. Even today I feel bad about that.

Vietnam changed us all.

Air Force Vietnam Vet: Brother Jack, (almost in tears) I left country on a "Freedom Bird" from someplace south as I served '67-'68 in I Corps DMZ Dong Ha. So-called 'retrograde amnesia' doesn't permit me to remember nearly anything. Docs here say my "homecoming" must'a' been traumatic. I immediately got a job (cop) and only now in retirement is it all coming back. PTSD every other week and here I am 70 years old. Something's wrong.

Army Vietnam Vet: There was that decade of "Crazed Vietnam Vets" during the 70s before PTSD was finally diagnosed for much of our problems. I always thought the media really loved playing it up. People killed people and they hardly ever made the news, but let it be a vet and there were films and follow-up, and the phrase "crazed Vietnam vet" was used at least 20 times per story.

Red Cross Donut Dollie Vietnam Vet: My first night home I woke myself screaming from a nightmare I do not remember. My father told my mother to let me be, I needed time.

Army Vietnam Vet: On leave for Christmas after coming home from Vietnam a few months earlier, I was home with my family but I felt alone. It took many years and many ruined holidays before I really came home. It got better after finally meeting and talking with other 'Nam vets in the early 90s. I still can't believe the person behind that blank stare is me.

Red Cross Donut Dollie Vietnam Vet: Sirens, a block away from my bedroom window, for the volunteer fire department. More than once I had to pull myself out from under the bed!

Army V ietnam Vet: When I came home from my first tour a neighbor's horn shorted out. I grabbed the wife and rolled to the floor.

Not sure where the grabbing her instinct came from, but had some "splaining" to do.

Navy Vietnam Vet: It was my first week back when the locals tested their sirens late one night. Hit the floor, much to my wife's surprise.

Red Cross Donut Dollie Vietnam Vet: I was in the park in Petaluma, California, reading, when the fire siren went off. I was under the bench!!

Army Vietnam Vet: First thunder I heard, went under the table.

Army Vietnam Era Vet: You could always tell the Vietnam vets in the motor pool in Germany when a truck would backfire.

Marine Vietnam Vet: Loud crashing sound, 4:00 a.m. Was that incoming or outgoing? A '63 Chevy hitting my '71 Datsun in Torrance, California.

Red Cross Donut Dollie Vietnam Vet: Last week I was talking to a new friend, a Vietnam veteran with mobility issues due to shrapnel, on the sidewalk on our little main street. Exactly behind him, a truck backfired. I dropped my bag, and he almost came completely out of his chair. We both laughed, and we both understood.

Army Vietnam Vet: I was just back from Vietnam in 1969, possibly three months out of the Army, and I was going somewhere with my sister in her car. The window on my side suddenly fell inside the door; I fell over on the seat. My poor sister almost wrecked her car, thinking I had been hit by glass. I explained to her the reactions we had to loud blasts or booms: "Get down!" or "Hit it!'
'

Navy Vietnam Vet: It was my first or second night home from 'Nam when the local volunteer fire alarms went off. I jumped out of my bed, instinctively reaching for my flak jacket and helmet. Scared the crap out of my brother who couldn't understand what was going on. I still jump at loud bangs and sirens.

Army Vietnam Vet: That "Thank you for your service" is lame. The only "welcome home" I accept is from another Vietnam vet.

Army Vietnam Vet: Heading home was special. What we experienced when we got back was a nightmare, from anti-war activists to the media. And from some prospective employers too. And nightmares. We never come back. One day in combat, next day home. Happy, but feeling guilty that we left our friends back there.

Air Force Vietnam Vet: Home in February '70. In less than a week I was ready to go back and be with the guys again. I felt guilty about something I had no control over.

Marine Vietnam Vet: I gave up the war a long time ago. It just won't give me up.

Navy Vietnam Vet: When I returned to college, my classmates seemed like they didn't want to interact with me. Most just ignored me or gave me strange looks. Thank God, there were a few other veterans there. I found it hard to finish, but I got my BSE. I never taught school, but carried mail 30 years. Vietnam still haunts me, and when I go to sleep I don't know if I will sleep here or in the Gulf of Tonkin.

Army Vietnam Vet: I recall sitting outside on my sister's patio in early June 1968. There was a party for me, my homecoming. At 11 o'clock at night I was having a smoke, all alone. I had this exact thought; "What the hell happened to me and how did I get here?" After that there were a lot more questions than answers.

Army Vietnam Vet: I'm still mad all the time—will not forgive and forget. Now, all I want is to be left alone for the rest of my life. I live alone, dine out alone, and travel alone.

Army Vietnam Vet: When it became time to DEROS (go home), I was conflicted, but a couple guys in my platoon said, "You made it this far, it's time for you to go home." Both of them had extended and I felt I should stay, but I took their advice and left. Never in my life, before or since, have I felt more alone. I felt I let my platoon mates down. Nobody understood and nobody seemed to care about what was happening so far away. Nobody asked me questions about it; they assumed that if it wasn't brought up it either didn't exist or would go away. After all these years, I think that attitude from my family and friends did more mental harm to me than the war itself, but I didn't know it until the last few years. I, too, shut down, as did a lot of the guys who served in Vietnam.

Army Vietnam Vet: It's been 48 years and I still hold bitter feelings, and I probably will until the day I take that dirt nap. I'm proud that I was able to serve and I'm proud of all others who did serve and are serving today. I hate and distrust the politicians.

Army Vietnam Vet: I was drafted. I was jerked out of civilian life and put through four months of intense training in preparation for the war. I came home on a 30-day leave. But it was out of focus. It was hard for me to enjoy the time, knowing that a war with my name on it loomed on the horizon. Then came the war. It'll toughen you up physically, mentally, and emotionally. It was a long, long year. Then it was over. I got on a plane, and 30 hours later I was standing in my mother's kitchen. In the next few days I realized that I didn't come home; instead, I returned to a country that resented me, almost to the point of hatred. And that's been the most difficult thing for me.

Chapter 8 - Marine Corps Stories

Marine Corps personnel were first sent to Vietnam as advisors. In March 1965, 3,500 US Marines landed to protect the base at Da Nang. The Marines performed perimeter guard duties and defensive sweeps of the local countryside. Shortly thereafter, they began offensive operations, seeking out the enemy to prevent attacks on United States' installations. An aggressive offense is the best defense.

By 1966, there were nearly 70,000 Marines in-country. Ground combat operations had increased from small unit operations to large-scale actions, searching for and destroying enemy troop concentrations and supply depots.

The Marines were assigned to I Corps where they would fight and die for the remainder of the war. I Corps was the northernmost area of South Vietnam. It was commonly known by the terms DMZ, Khe Sanh, Con Tien, Dong Ha, and many others.

The Marines had their own air arm in addition to ground combat troops. Helicopter squadrons evacuated their wounded, performed resupply missions, and inserted and extracted their troops on combat operations. Fixed-wing aircraft struck enemy installations and troop concentrations in South Vietnam, and targets in North Vietnam.

Fighting in I Corps was intense. In 1967 alone the Marines suffered 3,461 killed in action and another 25,525 wounded. Battles in 1968 during and after the Tet Offensive were horribly bloody. Khe Sanh and Hue proved to be some of the heaviest fighting of the war.

In all, an estimated 513,000 Marines served in Southeast Asia. The Marines lost 193 fixed-wing and 270 helicopters during the war, and 14,840 Marines died there, 51,392 were wounded requiring hospitalization, and another 37,202 were wounded that did not require hospitalization.[1]

1. Casualty figures from Defense Casualty Analysis System (DCAS)

I knew he was broken, but I just didn't know how broken.

Corporal Andrew J. Brocato

12th Marines, 1st Marine Division

(Narrative by Melody Brocato, Marine Veteran and Widow of Andrew Brocato)

The year was 1968; the States were in turmoil. Race relations, war demonstrations, killings, burning of cities—what a place to come home to. I was stationed at Marine Corps Air Station Cherry Point, North Carolina. MOS3122, Airfreight Operations. This was my first assignment. I manifested planes going to Gitmo, Miramar, and lots of other locations. I filled the planes with equipment, supplies, and the leftover weight was men, all headed for Vietnam, eventually.

By chance, I met a young Marine who caught my eye. A real punk, who said he was from Chi-Town. I too was from Chicago, but that was really all we had in common. I was new to the Marines but he was a "Salt," already back from Vietnam. He had been wounded, lost partial sight in his eye, shrapnel in his neck, and stepped on punji stakes that almost cut off one of his toes. He was bitter and hated the war. He was seven-and-a-half miles from the DMZ, and had lost many friends. He was with the 12th Marines, stationed at Camp Carroll. They had been overrun a couple of times while he was there. Eventually, the jungle had reclaimed the whole LZ. He was a truck driver and humped ammo for the 105's. He made many trips back and forth to Chu Lai and the surrounding villages. He was awarded the Purple Heart, National Defense Medal, Vietnam Service Medal, Vietnam Campaign Medal, Vietnam Cross of Gallantry, and the Presidential Unit Citation Award. They say he was one of the lucky ones: he made it home. But did he?

I never knew him before the war, so the only thing I knew was the boy I met and fell in love with. I knew he was broken, but I just didn't know how broken.

The Vietnam War had taken the life out of him. We married in 1969 and we began our civilian life in 1971. What followed was a series of fights with the VA about his service-connected disabilities. He was turned down over and over again. I felt he was a ticking time bomb.

His parents didn't know their own son when he returned home. The gentle altar boy who left came home a mad man. We did get him to see a psychiatrist. That helped a little, enough to get him stable and to be able to function as a husband and father. Every day was a challenge.

He became a police officer: he always wanted to "Protect and Serve," and that's why he joined the Marines. He still suffered from sleepless nights, and when he did sleep there were nightmares. He worked as a police officer for more than 25 years, always working nights, so as not having to sleep.

Three years ago we lost our 32-year-old son. He suffered many years from a chemical imbalance and was on many drugs to try to make him normal. Doctors think it may have been hereditary from his father's Vietnam experience with Agent Orange. He died of a morphine overdose, which was prescribed to him.

Eventually the skin diseases and internal damage from ingesting Agent Orange while in Vietnam took over Andrew's body. Surgery after surgery was performed without results. Ischemic heart disease couldn't be kept under control. On July 6, 2015, he died. The war never left him until that time. He has finally come home, but not to his children and me. He has come home to Jesus.

On October 4, 2015, his name was engraved on the Vietnam Agent Orange Memorial at Midway Village in Rockford, Illinois. The memorial is dedicated to those Vietnam veterans who are done fighting the war. Those Vietnam veterans, like many others, gave the ultimate sacrifice. They are the men who suffered all those years with the silent killer, Agent Orange. The Vietnam War has taken so much from all of us. His name will live on forever on that wall. His name was Andrew J. Brocato, USMC. My name is Melody S. Brocato. By the way, I am, and will always be, a woman Marine.

Corporal Michael George Munro
1st Battalion, 4th Marines, 3rd Marine Division

The Tet Offensive raged on from January through late February 1968. We pulled several ground operations and stood lines at Con Thien near the DMZ. On June 5, 1968, we went on operation LZ Loon. During the night, our lines were being probed, and it was obvious that an early morning attack was imminent. Even though we had the high ground, the NVA was soon in the perimeter. Rounds were continuously hitting our position from behind. It was then I was shot in the head. There was more damage from an RPG, and now things got life threatening. Lying there among other wounded and the dead in a bomb crater, I thought about my mom and did not want her to find out I was injured. Automatic and semi-automatic weapons fire snapped overhead from all directions. I always wondered how I would react should I ever get wounded. Now, what with the dry heaves and pain to my head that was untreatable, I knew.

I went in and out of consciousness several times, hoping that when I came too, the fighting would subside. Dried blood had closed my left eye, and a peek now and then was all I could muster. From the heroic efforts of others, I was literally thrown onto a medevac chopper and flown out of the area while under intense fire. I was transported to an underground triage bunker, and I could hear several voices I knew from my platoon talking while being treated for wounds. From there, it was a long chopper ride to the USS hospital ship, Sanctuary. After ten days, I was sent to Japan for an additional ten days before I was on a flight holding close to 300 wounded to Washington, DC. My final destination was Norfolk Naval Hospital in Virginia.

As far as my religious beliefs, there is no God. Vietnam proved that. How can I believe in a so-called loving God when all I have is seen is the horror of war, and contributed to the slaughter? I'm guilty of this insanity, because to do such terrible things one must get into the insanity mode or frame of mind. It would take me days to come down off that rush sometimes. I can truly say there was no hatred in me for the enemy. I killed them for self-preservation, and not because they were communists. A communist? What was that? I wouldn't know one if I fell over him on the street. Hell, I was almost nineteen by then. I was taught to shoot when shot at. How simple can that be, unless you are brainwashed?

On two different camping trips, I attempted to face my demons and go to the Vietnam Memorial Wall in Washington. The third attempt was successful, seeing the names I could remember cut into stone, like they were cut in my heart. Back then, I could not take the time to mourn, but, now, staring at that black granite was the right time. My wife back then knew it was my moment to reflect, and she remained behind, somewhat letting me embrace my moment of deep sorrow and guilt because they couldn't be here.

It's my hope that whoever reads this will get something out of it. I ruined my first marriage because of that damn war, and when she got religion and tried to force it on me and I was having no part of it. Save me? That wasn't going to happen. I'm no hypocrite, thinking I can accept Jesus as my savior. What I have done cannot be undone. Maybe I just want to be in a living Hell because I deserve it. Since getting out of the Marines, it's been two divorces, years of group therapy, counselors, and psychiatrists, not counting all the medication, and a psych ward. Then it was a motorcycle accident and I lost my leg. If that wasn't bad enough, now I have cancer, and I never have smoked and don't drink much.

Sergeant Douglas J. Gillert
US Marine Corps, Public Affairs Office, 1st Marine Division
SMSgt U.S. Air Force Retired

I enlisted in the Marines right out of high school in June 1967, and served four years, including a year in Vietnam as a combat correspondent (reporter). I mustered out from the Corps in 1971 with the rank of Sergeant (E-5). In October 1972, I enlisted in the Air Force and served in a variety of assignments, mostly as a writer and editor, retiring in December 1991 as associate editor of Airman Magazine, the official Air Force magazine, with the rank of senior master sergeant (E-8).

I joined because I believed in what John F. Kennedy said at his inauguration; "Ask not what your country can do for you—ask what you can do for your country." On a more practical side, it was the only way I could find to get out of Michigan, a place I didn't leave until I boarded a 707 and headed for Marine Corps boot camp in San Diego. After a couple of years outside, I went back in because I just couldn't find a place for myself back home. I had PTSD but didn't know this until just a couple of years ago when it really took hold. As long as I was in uniform, my work life, my professional life, was great and I excelled. I struggled through and eventually sabotaged three marriages, wrestled with drug and alcohol abuse, and adultery.

I volunteered for Vietnam and was happy when I got my orders. I spent two-and-a-half years at Camp Pendleton, and by this time I felt I needed to find out for myself about Vietnam. I was fully against the war but that didn't matter, I still needed to see if for myself.

When I arrived in Vietnam, I was assigned to the Public Affairs Office, First Marine Division, in Da Nang, but mostly worked out of Landing Zone Baldy with the 7th Marine Regiment in the Que Son Mountains, Que Son Valley, and Antenna Valley south and southeast of Danang.

Coming Home

I was selected by the skipper in our office to represent Public Affairs in a contingent of Headquarters Battalion Marines officially returning the unit colors stateside, as the Marine drawdown sped up in the latter part of 1970. As such, I came home on a naval ship, the USS Juneau City. This was a 27-day journey from Da Nang Harbor to Long Beach. When we arrived in Long Beach, we were so excited to see all the beautiful, round-eyed women in bright sundresses, halters, and shorts there to greet the arriving flotilla of three ships. There was even a band playing. We thought it was for us, but soon learned they were all there to welcome their sailors home. They didn't even know we were aboard, and we were snuck down a rear gangplank out of view, rushed onto a bus, and whisked away to Camp Pendleton. We were told they moved us out this way because they didn't want to cause any trouble.

The trip home was surreal. Nobody—NOBODY—even asked me about Vietnam. Not my mother, not my father, not my sister, even though her fiancé served in Nam before me, and he didn't even talk to me about it. Or maybe he did and I blew him off. I found that I suddenly didn't want to share that I'd been there, that somehow it wasn't a good thing to have done. I kept quiet about it.

I got a leave and went home for the holidays and then returned to Camp Pendleton for three more months of active duty. I left the Marines for good to enter college in March 1971. I made the best of it. I didn't stay in college, instead going back to work for the menswear company I'd worked for during high school. When they opened a new boutique up the street, they made me manager, but the pay was not good and after getting married I started looking for better opportunities.

Nobody ever mistreated me. But, again, nobody even asked what it was like. Was I sending out some radar signal not to talk to me about the war? I don't know. It wasn't much different when I went in the Air Force. Nobody wanted to know about my Marine Corps experiences. I just buried it all inside.

Home

Serving in the Air Force for 19 years, I gained a bit of recognition and gained a lot of respect from my family, but I no longer had any of the friends from my past, just ones I'd met at work in the USAF. I excelled at what I did and received many awards and accolades over the years, while inside, I continued to smolder and burn. It sometimes felt like I was slowly devoured from some internal and endless well of burning acid.

I had a slew of problems afterward. They all had to do with an inability to establish a trusting relationship with my spouse(s). I didn't attribute any of this to the war until recently, when through therapy I was brought to discern the true emotional and psychological impact of my combat experiences.

I started seeking help when my last wife and I were struggling to save our marriage, which we did not. I continued seeing a therapist and then switched to another and have been going since. I also went to the VA and was awarded 60 percent disability for combat-related PTSD this past spring.

My wartime experiences profoundly changed me. One thing is certain: it created within me a nearly constant sense of unease that only goes away when I can completely absorb myself in something to the extent I forget for a while.

Looking back, I feel wiser, sadder, stronger, but colder, too, and sometimes a little unbalanced.

Sergeant Wayne Stafford
H&MS-36
HMM-265

I was burned out on school by 1966: actually, by 1963. Just getting a job wasn't an option because of the draft, so I decided to enlist in the Navy. I also didn't want to miss out on our only war at the time. This is logic according to an 18-year-old. I drove to downtown Concord, California, to join up. With my dad 20 years Navy and retired, it made sense also to join the Navy. It was 1966, with the government drafting everyone with a pulse, so why not? While talking to the chief, the phone rang. After 15 minutes of him jaw jacking, I walked out and saw a guy in a cool uniform standing next to a poster with a picture of an airplane in the lower right corner. "You got airplanes?" "Yup," was his reply. "What do you do for a living?" I asked. "You'll find out soon enough." Done.

I entered MCRD, San Diego, September 24, and graduated November 21, 1966. I received my draft notice while in boot camp and asked the drill instructor for guidance. He said, "Throw it away."

I finished ITR just before Christmas break. From there it was six months of Aviation Electrician "A" School. followed by a year at El Toro. I received another draft notice while stationed at El Toro. My OIC said, "Throw it away." Fast forward to 1972, and I called my draft board and asked for my new draft card. After a short delay on the line, the lady exclaims, "Where have you been?" I answered, "Vietnam, Thailand, Philippines, etc." Her response? "Never mind." New draft card arrived two days later.

After I left El Toro I spent three weeks at Staging Battalion, Camp Pendleton, eight months at Phu Bai, five months on the USS Iwo Jima doing operations off Vietnam, and then finished up at MCAF Tustin (LTA), California, in 1970.

My MOS was an aviation electrician for transports and helicopters. I went in the military because I wanted to work on fixed-wing aircraft. I built all the models as a kid, with the F-8 Crusader being my favorite. Helicopters were interesting, but I considered them somewhat of a novelty. The closest I had been to one was an H-19 landing on the golf course across the street from Kenny Weaver's house at Pearl Harbor when I was seven. In 1964, a Sikorsky UH-34D roared over the ridge above our heads in Irvine, California, when Kenny and I were firing his M-1 Carbine with tracer rounds. Scared the crap out of us. Little did I know, less than four years later I would take my first ride on a helicopter, and it would be a UH-34D. Fixed-wing was boring. Not enough moving parts.

I was apprehensive about Vietnam, like everyone else, but when you train as a Marine, it's part of the deal. While stationed at El Toro we had about two guys a month get orders for Vietnam. You would get out of "A" School, spend some time in a squadron learning the ropes, and then you were gone. Spring of 1968 I figured my time was coming, but I had just started dating this little Southern California beach beauty, so I didn't want to miss out. A deal was made where if I volunteered for Vietnam they would delay my orders until the summer was over, so I reported to Camp Pendleton in late August. Got a Dear John anyway.

My squadron at El Toro had a mix of aircraft, and was basically a transport squadron except for the two UH-34D helicopters at SAR. I was assigned to H&MS-36 at Phu Bai, because they had both UH-34D helicopters and a C-117 like those at my previous squadron. The primary job of H&MS-36 was second-level maintenance on helos that had issues above the squadron's capability, but I worked the flight line and didn't get involved in the normal H&MS functions. In December 1968, I started training to be a helicopter gunner, and flew combat missions part time when the boss let me get away. HMM-363 went home to rebuild with the CH-53D in January, so they gave us a whole gaggle of H-34s and crews. This really kicked our flying combat missions into high gear. Around April 1969, the UH-34D began the transition to the Reserves so H&MS wouldn't have any more helicopters, so I transferred to a CH-46A squadron and began the process of learning a new aircraft. I arrived at HMM-265 on about April 23 and immediately went on night crew. We left Phu Bai and moved onto the USS Iwo Jima June 6, 1969.

In September 1969, the sergeant promotions came out. I had spent a lot of time as a lance corporal with promotions frozen at El Toro, so I missed the cutting score by a fraction. The CO wanted two more sergeants, so he asked for those with combat aircrew wings. I was sitting in the right seat of a CH-46A on the flight deck of the USS Iwo Jima when our squadron top came up and told me I made it but don't tell anyone. Later, while walking through the ship near the ready room, a hand came out of the admin. office, grabbed the front of my jungle jacket, and pulled me in. I was handed a Sergeant Warrant, and my new chevrons were "pinned on" by everyone in the office, and I was given three days at China Beach.

Coming Home

I caught the first helo off the USS Iwo Jima in mid-October 1969, and immediately headed for the transit barracks at Da Nang. Unfortunately, my arriving early made no difference in my departure time, so it meant another night in Vietnam. Three things I remember about spending the night there. First was saying the words, "Going home." It was almost like a foreign language, having not spoken them in over a year. Another was in the head, sitting on the can reading all the graffiti on the inside of the stall. I almost split a gut laughing. It was all "in-country" jokes which you had to be there to get. Finally, the sea bag inspection. The only thing issued in-country that I was allowed to keep were my jungle boots. The next morning we caught a plane for Okinawa. We were rejoicing when the plane lifted off. What relief.

Upon landing at Kadena, we were hauled over to Camp Butler for processing and to pick up the stuff we left behind and did not need in Vietnam. I ran into three guys from my staging battalion squad at the deli, and what a change. At Camp Pendleton, they were three kids fresh out of ITR/high school, so being their squad leader was more like daycare. Now, they had aged 20 years, after spending 13 months humping mortars. My last day of processing was medical, which I had just finished and was walking through the building on my way out when I heard helos. As I stepped out into the sunlight, the CH-46s from my squadron on the USS Iwo Jima were landing on the baseball field across the street, dropping off all the guys who were short. The ship had sailed from Da Nang to Okinawa on the way back to CONUS.

That afternoon we boarded a 707 for home. The plane departed about 1700 for a direct flight to MCAS El Toro. It was quiet the whole way. The first sight of CONUS was Monterey Bay, with arrival about noon. The parking lot at El Toro had about six cars to welcome 165 Marines, so most guys were directed to different modes of travel, either air or ground. I was lucky. My mother lived about 40 minutes away, so I called her to pick me up.

Mom showed up, I jumped in the car, and we were gone. Drove down to the main gate, and there was a welcoming party awaiting us. Crowds of protesters were yelling obscenities at all who passed. We ignored them and continued.

As we pulled onto I-5, we approached an overpass. Keep in mind, I spent the last year plus with helicopters, and hadn't been under anything other than a tree limb in all that time. I began to pull up on the car seat wanting to go over the obstruction. Another thing I noticed during the ride home was my mother had aged ten years. She isn't one to show emotion, but I think my being in Vietnam had taken its toll. Then she asked me, "Who's this Topper guy? He keeps calling to ask if you're home yet." Topper was with 3rd Recon, whom I'd met on the ship. Both being from Southern California, we hung out until he rotated a month before me.

I called Topper when we got to mom's apartment, to find out he had set up a double date for that evening. We were going to the movies with the "Twins." Later that evening he shows up in a 1961 VW van with homemade camper top, and it's off to pick up our dates. We arrived at the movie about 6:00 pm, but a little late, so we sat in the front row. I spent most of the feature looking around at girls and people not in jungle utilities. Quite an adjustment. To this day, I have no idea what the movie "If" was about. Over the next two weeks, I noticed a mental change. The best way to explain it would be two scenarios. In-country Attitude: Cruising along in a helicopter when an NVA opens up with an AK. My response? Kill the mother f***er. Stateside Attitude: Guy jumps out with an AK on Crenshaw Blvd. and opens up. Response? Dive for the curb and shit my pants.

I spent most of my time around Topper's folks' place and hanging out with friends. Discussion of Vietnam didn't come up except between us jarheads. I didn't get called a baby killer until 1976. Pete Werback had started a company called Schmegg, building high-powered audio amplifiers for cars. It allows you to vibrate your car and all those within a four-block radius. I was helping him stuff some PC boards one evening when Pete asked me if I had served in Vietnam. I answered yes, and his girlfriend (later, wife) said: "So, you're a baby killer." The company was later renamed Linear Power.

Being a Navy family, we moved all the time, so I didn't have a huge collection of hometown friends. I drove up to Northern California to visit a few, with no talk about Vietnam. My dad and his family had moved to Greenwood, Arkansas, and he didn't even know I had made it back. I didn't see him for at least another five years. I have been to a couple of high school class reunions and still don't know who those people are, even with a yearbook. One of our high school classmates maintains a website to keep people connected, and he's acknowledged those who served and those who didn't return. It's appreciated.

I finally decided to visit my dad in the mid-1970s. Keep in mind, my parents divorced in 1960 and my dad was remarried with two kids, so I was more of a visitor than family. My step-mother was only 14 years older than me, so she didn't know what to do with a teenager, and I didn't know what to do with a 20-something. We're close now that we've grown up, and dad is gone.

My father didn't ask about Vietnam, and I wasn't interested in talking about it anyway. Wasn't comfortable when I lived there, so still no comfort. My father stopped answering my letters after three months in-country, and he moved while I was gone and didn't tell me where. Remember the joke about the kid coming home from school and his family had moved?

I made new friends, and my family doesn't talk much. I only communicate with one classmate from high school, and one from the high school I would have attended had I not moved. Eva attended the school where I graduated and lived across the street. Her son is a Marine Cobra driver now, stationed at Quantico. Nancy and I went to school together from fifth to ninth grade. We connected when I was talking to the wife of a boot camp, ITR, "A" School, Vietnam buddy of mine. Turns out her best friend in high school was my pal. Nancy and I try to get together at least once a year and hang out for a couple of days.

Home

My family invented dysfunction. I made new friends who didn't know I was a Marine in Vietnam until I joined Popasmoke and reunited with some jarheads from back in the day. They are my family. Now, Vietnam is in the past except in my head.

I was living in northern California in 1975 when Saigon fell, and my first thought was, "What a waste. All those Marines lost for no reason. We fought for I Corps, and that was the first place given up."

Years later, in the mid-1980s, I went through a period of survivor's guilt. It just hung over my head like a cloud until one day I thought about a good friend, Jim Hannibal. Jim and I went through boot camp, ITR, "A" School, and west coast duty stations together. We even had racks across from each other at Phu Bai. Jim was KIA January '69 during a night medevac. The thing that got me through this episode was thinking of what Jim would say if he knew about all this guilt I was carrying around. It would be something like, "Get your head out of your ass. It could be a lot worse." It worked.

I went to the VA Agent Orange screening where the doc said I'm the healthiest candidate he's ever seen. Good news so far. Every year my lab results are in the green.

I appreciate what I have and have a real short temper with people who don't. It's also comforting to know how I handled being shot at. Kind of a guy thing. I found out I'm a lot tougher than some tough guys I've met over the years.

In early 1997, I was visiting my dad in an Arkansas hospital. During a lull in the conversation, I wired my laptop into the hospital phone system and checked online for any Navy reunions for my dad. I saw the Popasmoke website, so made a mental note. When I got home, I checked out the site and saw my old OIC was a member. I joined, and attended a squadron reunion that summer. I really enjoyed it. One weird situation was seeing John "Ace" Hunt walk on crutches into the meeting room. Last I heard of him was when I was at the Marble Transit Hooch waiting for a hop back to the ship. Word was, Ace had fallen out of his helicopter and was medevaced. Was good to see him. Also, anyone who says the fall affected his brain is wrong. He's always been that way.

There was no change in my faith, but I sure am a believer in guardian angels. When that 122 hit next to the Charlie taxiway less than 100 yards away, it got my attention. That, and a few other episodes.

The Traveling Wall came to Reno, Nevada, while we were visiting in-laws, so I had to see it. My reaction to seeing all the names was as expected, but the real shock came when I saw Jim Hannibal's name. To see the name of someone you knew so well was different. I've since been to the Wall in DC, Pensacola, and Jacksonville, North Carolina.

My feelings are a conflict of thinking, "Should we have been there at all?" and, "Someone had to poke a stick in the eye of communism." The fact that we won militarily and lost politically. Seeing our elected officials at their worst, abandoning South Vietnam. On a personal level, I remember many times standing on the flight line or cruising along in an

H-34, thinking this is the biggest adventure I will ever experience, and it will all be over before I'm 22. At least I experienced it with Marines. America's best at their best.

I made a deal with God in Vietnam that if I made it home alive, I wouldn't shoot at anything that didn't shoot at me first.

Sergeant Stephen Simonton

Delta, 1st Bn., 7th Marines, 1st Marine Division

This is the story of Stephen (Steve) Simonton. Much of this story was written by Steve before his death. The rest was compiled from recollections of his wife, Sandee, as well as from copies of letters written to the VA in regard to Steve's PTSD, by his daughter, Erin, and his sisters, Kim and Pam.

Steve was born in Miami, Florida, March 5, 1948. His mother was a war bride, born in Ireland but living in England in WW II where his dad was stationed. His older twin brothers were born in England.

Steve came from a background of families that served their country in times of war and peace. His father and uncle both served in the WW II. His mother worked as a civilian in a bomb factory in Coventry, England. Her older brothers served in the British Army.

His older twin brothers joined the US Air Force. Soon after, in 1966, at the age of 18, Steve joined the Marine Corps. He felt that it was his duty as a young American to serve our country in its time of need, just as his father, grandfather, and great-grandfather had done before him.

It wasn't long before two of the three brothers found themselves in 'Nam. One served in administration, seeming to avoid much combat, especially after contracting malaria. Steve saw only combat his entire tour of duty, mostly as a radio operator, heading out into the unknown with the battalion. The family lived in constant fear of losing the boys to the cause.

Steve arrived in Vietnam at Hill 41 just days before Christmas 1967. It was just getting dark, and he was assigned to first squad, second platoon. He went on his first squad-size ambush that night.

Delta Company's mission was part of the first line of defense perimeter (approx. 20 miles) around Da Nang and its airstrip. The enemy came down out of the mountains into this area to get within range of Da Nang with the rockets they had. Delta ran squad ambushes two nights, then a day patrol the third day. Their night off was bunker watch on the perimeter of the compound. Some nights in the bunkers he could see a firefight in the distance and wonder which of his units had stepped in the shit and who wouldn't be coming back in the morning.

He took over as squad radio operator after a couple of months in-country. Radio was their lifeline to the base camp when they were out in the bush and needed fire support or a medevac. Sometimes they would go for a couple of weeks with little or no action. Other times it seemed like they couldn't get off the hill without someone tripping a booby trap or getting into a firefight.

Operations were the worst. When the monsoons stopped, they ran operations from a platoon size to multi-battalion size. The bigger the operation, the worse the beating they took. On one operation, Steve was radio operator for his platoon commander, Lt. George Coleman, one of the best they had. Second Platoon was the lead element in the assault of the enemy base camp, when Lt. Coleman gave the order to advance. As soon as they stood up to move forward he took a burst of automatic weapon fire across his chest. He was killed instantly. Steve could hear the rounds whizzing past his head. They stayed pinned down until the air support arrived with 250-pound bombs and napalm. You could hear the bombs clicking as they were released. They call it close air support. Nine of the eleven guys in his squad and many others were killed or wounded.

Eventually, he was moved up to company radio operator. He and another Marine took six-hour shifts in the commo shack, monitoring the units in the bush. He only had to go in the bush when the whole company went.

Coming Home

(From a letter Steve wrote to the VA)

My last medevac was after we moved down to Hill 37. Don't remember his name. Had his leg blown off above the knee. We loaded him and his half a leg with the boot still in place onto the chopper and sent him off in the dark.

Next morning, I got my orders: I was going home. Four or five days later, I was standing in my parents' living room. Back in The World, 30 days' leave. Hot showers, clean bed, real chow, not getting shot at, all the things I'd craved all those months, weren't so great. I was bored before the 30 days were up: no action.

Back at Camp Lejeune, I was sent out to Camp Geiger as an Infantry Training Instructor. Having been there as a trainee 15 months earlier, I remembered thinking then a lot of it was bullshit. All the instructors were highly motivated. We knew what we were sending them into. I often wondered how many of them wouldn't make it back.

August 28, 1969, I was released from active duty. Early release. Didn't have enough time left to send me back to 'Nam.

I married my high school sweetheart, went back to work for my dad's plumbing company. Back to the real world to get on with my life. My new bride soon realized the person she fell in love with was not the person who returned from Vietnam, and the union was a short one.

Steve's sister Kim recalled how Steve was after he returned from Vietnam:

"I was in high school when Steve enlisted and went to Vietnam. I remember when he returned from Vietnam, he seemed to be a shell, not a person. He would not engage in conversation, and never spoke about what he did or saw while he served in Vietnam. It was evident from the beginning of his return how different he was. When he slept, it was with his hands wrapped around an (invisible) M-16. You had to stand across the room from him to wake him so he wouldn't attack you. Any loud noise was disturbing, and he would be looking for cover.

"As he saw how these things upset everyone, he withdrew even more. He no longer chose to be around family. He was not able to cope with social situations outside the family, and withdrew from contact with everyone. It was difficult for him to concentrate on what was going on, and he had a hard time connecting with anyone.

"He basically removed himself from the family and had little contact with us. It was as though he felt he had nothing to offer. His experiences in combat were so horrific for him that he could not relate to what we called "normal." The experiences he had had a serious effect on his family connections and his ability for social functioning.

"The lack of support from the public when Steve returned from Vietnam really made him question what he held to be his beliefs: to help others, to be considerate, to serve, and to have honor. His experience made him question who he was and what he believed. His inability to connect with his family and friends and the lack of services provided have had a lifelong impact on him, as well as on his family."

Steve's sister Pam wrote about the changes in Steve after he returned:

"When Stephen returned home he was unrecognizable. He had lost a lot of weight, something he could ill afford at the time, and he moved like a ghost, never quite comfortable with his surroundings. We quickly realized it wasn't a good idea to accidentally startle either one of the boys, as their reaction was to strike out to protect them from some unknown harm.

"Stephen was much more withdrawn, detached, and unemotional than his brother. He had little to say about 'Nam. It appeared he had been in one long nightmare. Why did he return without physical injury when all but five of the 250-member battalion were injured or killed? Some of his buddies had been blown to pieces right beside him.

"Of the few horrors he was able to talk about, the most chilling was the description of the villages being overrun by the battalions. Most often, only the women and children remained in the villages, and they were slaughtered, something his nature and upbringing had ill prepared him for.

"By the time the brothers returned to the States, the disdain for the Vietnam War was well underway, and the returning soldiers took the brunt of this disdain. They were ridiculed, taunted, and spit on. All his values and beliefs had been totally undermined, and no support was forthcoming from the country he served.

"For many years, he existed on the fringes of life, distrustful of engaging with family and friends. It has been a long journey trying to find his way back, the war having changed his life forever."

Home, a narrative by Steve's wife Sandee:

Steve related to Sandee his thoughts about being a Vietnam veteran:

"The whole country seemed pretty much into the anti-war movement. We were baby killers. I found that offensive. What the hell happened? I grew up in the WW II era: Dad, uncles, John Wayne. A veteran was a good thing."

Steve got involved with the Vietnam Veterans Against the War. Planned a peaceful demonstration at the presidential convention in Miami Beach. They were told if they didn't leave, the National Guard would be called in to remove and arrest them. Who had a better right to protest the war than the men who'd been there? Calling in the military to arrest veterans? That made him very angry at our government and the way they were treated. He found out he was being investigated by the FBI as a radical. He drifted away from VVAW; it wasn't helping him, anyway.

He left his wife and his dad's plumbing business and drifted for the next 20 years: New York, Michigan, Tennessee. He went to the VA Hospital in Nashville, Tennessee, for treatment of PTSD in 1981. His worst memory was leaving his communication post after his shift, and a young mother and son with their cow (who was grazing) ventured too close to the perimeter. He heard the radio transmission calling for fire and tried to communicate to the radio operator on shift that they were villagers and no threat. He was too late. The mother was killed. Steve went down there and the son (four or five years old) asked him why he had killed his mother. That memory tortured him for the rest of his life. When My Lai hit the news, no one wanted to believe Americans did that stuff. Steve knew it could happen. Steve felt My Lai was the reason they were called baby killers. It appeared he felt guilty that he was a part of it.

For four months he was in group sessions. Everyone was trying to be crazy to get a disability rating. He didn't want to be disabled, he didn't want to be crazy. Group wasn't helping. He couldn't find a job, couldn't afford the gas to drive back and forth to the hospital 45 miles each way. He quit going.

Steve met Sandee in 1987. His second wife had thrown him out and he was living in the family plumbing business warehouse. His life's possessions were in a 1973 Ford van in the Hefty trash bags his ex-wife had so kindly packaged them in. Sandee had no clue he was a veteran, a Marine, nothing; he just didn't talk about it. In fact, he didn't reveal much of anything, including his current living situation. He did say he was separated, but details about himself took a long time to draw out.

Memorial Day weekend the "quiet" guy she'd become accustomed to became absolutely withdrawn and sullen. She immediately took it personally and thought she'd done something. He finally explained it wasn't her, that Memorial Day was a tough time for him, that he'd lost a lot of buddies in Vietnam. Revelations about his service and time in Vietnam were slow to come, but more revealing as he came to trust her. She still didn't know everything that happened during his service, but what he told her was a living Hell that no one could've lived through without emotional scars.

Steve had trouble expressing himself. He spoke slowly, in fractured sentences. He didn't communicate with his family. If they called, he would speak with them. He comes from a large, tight family, but he would not make an effort to communicate even simple things. To most people, he is totally anti-social.

Steve was quiet and depressed most of the time. He could not talk about Vietnam without breaking into tears, and then he would retreat away from everyone until he was feeling better. He wasn't comfortable in any type of group situation, including his family. He found it difficult to show love or anger; he simply would retreat into himself.

Steve's daughter Erin noted:

"At the age of 41, he became my father. Growing up, I had always wondered why he would awaken so easily at the slightest sound or movement in the room, why he could not make it through the first ten minutes of a Vietnam War documentary on television without tearing up. I would wonder why he was so against keeping fireflies we had caught in a jar, but he would always say, "If you keep them, they are just going to die. Why kill something for no reason?" I loved going through his photo album, always asking questions and wanting to know more, but I always felt like there was so much he was not telling me. I will never forget a picture of a fellow Marine with a bandage on his hand, holding up his hand that was bandaged and missing a couple of fingers. However, he sported a large smile. My father pointed to this picture once and said, "You know why he looks so happy even though he has lost his fingers and was in excruciating pain? He knew he was going home." Posttraumatic stress disorder was something that my father had always joked about, but we all knew that this was something he was suffering from".

By 1993, Steve and Sandee had married, had three daughters in three years, and received an invitation to a Delta 1-7 reunion in Washington, DC, for the 25th Anniversary of the Tet Offensive. He had mentioned that he'd like to see the Wall but didn't know if he could deal with it. Sandee literally forced him to go. He finally agreed when she told him that they'd go to New York City and take the train to DC on Friday for the opening cocktail party. They'd go to the cocktail party, and if he wasn't comfortable, they'd get back on the train Saturday morning, and go back to her friends in New York City: a second honeymoon, with his sister watching the kids. He agreed, on that basis.

Amazing! Reuniting with the "guys" and doing a special memorial service at the Wall with them was the best healing any doctor could've ordered. The man of few words thanked Sandee on the flight home for asking him to face his fears. (Yes, they stayed the whole weekend in DC.) He said that he'd spent 25 years dwelling on the dead, and that reuniting with the living who had survived the same experience had been the best medicine he'd found so far.

But Steve wasn't out of the woods, far from it. Two of his brothers had died, one by his own hand after many years of drug abuse, the other basically drinking himself to death. Steve had to go to the apartment where his brother had committed suicide. It was Thanksgiving morning. Sandee refused to let him go alone; she was afraid he might ultimately hurt himself if he did. It was the most awful scene she has ever witnessed. Keith had shot himself, and the body was there a week before anyone discovered it. Sandee had never experienced the "smell of death." Steve stoically did what was required, and left as fast as he could. He never said a word that whole day, and was depressed for months after. He later told Sandee that you never forget that smell as long as you live.

When the second brother died, Steve took it especially hard. They had been close, and Sean had been staying with us up until two weeks before his death. Steve had offered to help him find a place to live. He had come into some money from a settlement and had been staying with us for almost a year. Sean never said goodbye, he was just gone that afternoon when Steve came home from work. He was found dead in a motel room two weeks later. Steve never did get over the guilt of his death. The survivor's guilt reared its head again.

In 2005, Sandee and Steve bought property in South Carolina and planned to retire there. Twenty acres of woods, with lots of wildlife; deer, etc. He was adamant that he wanted a large property with no close neighbors. Sandee's father was a hunter, and she mentioned that she hadn't had venison in years. While he expressed that he had no problem with someone else hunting on the property, he explained that he would not, even though he had gone hunting as a boy. He said, "I made a deal with God in Vietnam that if I made it home alive, I wouldn't shoot at anything that didn't shoot at me first."

Steve retired in 2012. They prepared to sell the house in Florida and start building their retirement home. Steve was treated by the VA for a recurring earache on more than three occasions before he became totally unsteady and off balance. By the time they did a CT scan, he was in a wheelchair. They found six tumors on his brain. The cancer had metastasized. He had been under the constant care of the VA for over three years at that point, and no cancer diagnosis had ever been made or even looked for. He was then diagnosed with lung cancer.

They started construction of the new house in 2013, hoping it might be complete before he died. He didn't make it. Steve died in Fort Lauderdale on August 21, 2013. They were able to have his memorial in South Carolina, and placed his ashes and a marker on the property.

Chapter 9 - Dealing with the Effects of War

Time slowly ticked by for us Vietnam veterans. We struggled to adjust while staying in the shadows. We went to school, went to work, and some re-enlisted, choosing to stay where they were comfortable: in the service.

No one talked about Vietnam much after 1973. The protests stopped; we all started getting on with our lives. And then it happened. December 13, 1974, a probing attack was launched against the provincial capital of Phuoc Binh in the Central Highlands. It was now plain that North Vietnam would not observe the Paris agreements. Phuoc Long province fell, exposing South Vietnam's lack of resolve and military weakness. It also showed that the United States would not intervene.

General Van Tien Dung, second in command to General Giap, proposed an offensive to capture the Central Highlands in 1975. His plan was to conquer South Vietnam in 1976. This plan was approved and he took command of what was to be called the Ho Chi Minh Offensive.

The offensive began on March 10, 1975. I am sure that most Vietnam veterans were paying close attention to the news on their televisions and in the local papers. The biggest disappointment was that we, the United States, never intervened as we had agreed and promised. Our government turned its back on the people of South Vietnam.

I remember how angry I was watching the whole thing unfold. Town by town fell to the NVA. I watched my old base at Phu Loi fall. Unbelievable. Tragic. I was beside myself. I cursed, I was enraged. Every day I watched. Every day I became angrier and angrier. All the lives lost. All the wounded. All the civilians, dead.

The end came quickly, too quickly. Province after province fell with lightning speed. The noose tightened around the remaining troops protecting Saigon. A gallant defense was made at Xuan Loc, but eventually that fell too. The enemy entered Saigon. Finally, on the morning of April 30, 1975, North Vietnamese tanks and infantry entered the city and crashed through the gates of the Presidential Palace. The war was over, over for all the Americans who served there. Or was it?

We spent our youth in Vietnam and it was all going down the drain. For what? We were treated like killers and drug-crazed animals by the public and media. We were sprayed with Agent Orange, destined to die an early death plagued by cancer, heart disease, and a myriad of other illness. And our government just let it all fall apart. They abandoned the Vietnamese people and they abandoned us. This was our ultimate Welcome Home.

Army Vietnam Vet: I still don't feel welcome. I honestly feel animosity for the people who thank me for my service. I needed that in '69 when I came home. Got into a fight in the airport in San Francisco and, really, never recovered.

Army Vietnam Vet: Being unarmed during a firefight is a recurring nightmare of mine. I had no idea others experienced my worst fears.

Army Vietnam Vet: The end of Saigon—truth is I cried like a baby. Watching TV with my then wife of three years, she asked why I was crying. I told her something I never told her in the five years we were together. Never told her I was a Vietnam Combat vet. Americal 1967-68.

Army Vietnam Vet: When I heard about the fall of Saigon, I destroyed my house. Anything not nailed down got thrown through the TV or a window. Anything I could not throw got tumbled through the house until I collapsed in a corner.

Army Vietnam Vet: When Saigon fell, I felt physically sick to my stomach, sick in my heart, numb in my head. "WTH did I give 18 months and half my sanity for?" was all I could think. All the lives, all the widows, all the fatherless kids and lost brothers and sons. What for? What for?

Red Cross Donut Dollie Vietnam Vet: When South Vietnam fell, I felt so sad and so angry, and worried about those in Vietnam and what horrors they would be facing.

Army Vietnam Vet: My repeated nightmare for at least a decade was that MPs showed up at my house to tell me I had been discharged too soon and that I had to go with them to go back to Vietnam and do more time. My wife and kids were tearing at me not to go and I was clinging to them to keep from being taken. I'd wake up screaming and crying. My wife divorced me after 11 years. That made the MPs coming for me a cakewalk, ending that nightmare.

Coast Guard Vietnam Vet: My own siblings had no idea I had served in Vietnam until about five years ago. I was getting some paperwork out for a claim, and had my ribbon bar sitting on the sofa next to me, when my brother walked in. He saw that and asked about it and I told him. Needless to say, he was shocked, as were the rest when they found out. I came home on leave in uniform, but I guess they never paid much attention to it.

Army Vietnam Vet: I have only been in the woods with a gun once since I returned from 'Nam. I hit the ground three times in the first fifteen minutes. Never again.

Army Vietnam Vet: I lost 23 from my hometown, seven from my high school class in 1966. Living in my home town until last year, I was reminded every day of my guilt. Many times I wished I had died, even tried myself. Over the years I have learned to live for them and take care of veterans in their names.

Navy Desert Storm Vet: My dad served in Vietnam; I was in Desert Storm. Some years later I was taking a course at a college. I was wearing my ball cap. I had a young girl ask me if I liked seeing innocent people killed. My mouth dropped. All I could think about was my dad. Thinking about him going through this. My next thought was anger, and throwing her down a flight of steps.

Marine Vietnam Vet: Goddamn war turned me into a drunk and put my wife through hell, and almost ruined my life and marriage. Fifteen months of my life, wasted. I was only 19 years old.

Army Vietnam Vet: I learned two lifelong lessons in Vietnam. The first: I am who I am, take it or leave it. I have no need to impress anyone, for I have faced the devil and I am still walking the earth. The second, learned during monsoons: you can only get so wet. Once you get that wet, you just can't get any wetter.

Army Vietnam Vet: Backfires and sudden noises still affect me, and when my wife and I go for hikes, I have extra vigilance when we come to an "L" bend in the trail or other likely ambush sight. Years ago, I started wearing a Vietnam veterans cap and people would give the standard, "Welcome home." If they were my age I would ask them where they were when I came home and how did they treat vets then? Now I have mellowed and wear the cap because I am proud of my service to my country.

Sister of an Army Deceased Vet: My brother never discussed Vietnam, even if we asked. I think it would have been better for all if he had shared.

Air Force Vietnam Vet: Those who have never experienced war will never understand. You can't make it go away, it won't leave you.

Army Vietnam Vet: You know, recently I was sitting in a conference room along with my coworkers, and one guy I had respected as a boss blew me away. Somehow the Vietnam War came up, and he had gone to Canada. He said there was no way he was going to be involved in that stupid little war and has no respect for anyone who had! Just like when I got home over 40 years ago, I kept my mouth shut. I felt I did something wrong and then I also felt cheated. Here is a draft dodger and he is a head boss. Makes more money than I and is well respected by everyone in the room and company, and I have to answer to him about my department! I had had enough, and I retired. He never came to my retirement or said one thing to me before I left. Sad. Nothing has changed.

Marine Corps Vietnam Vet: Coming home from 'Nam, Sundays were rough, especially Sunday mornings. After spending more than a year over there where there was no God, we returned to where God was present, but it seemed hard to talk with God when we returned, for a long time.

Air Force Vietnam Vet: I came home from 'Nam and was warned not to wear my uniform. I still got spit on and called a baby killer . . . must have been my haircut and my duffel bag. Twenty-one years later, I came home from Desert Shield/Storm with people cheering, clapping, fire hoses going off, slaps on the back, and handshakes. I broke down and cried. It was a long wait.

Army WAC Vietnam Vet: Our nation is just waiting on us to die, and thus the story of Vietnam will no longer be with us. So sorry, guys, it feels like the very truth here.

Army Vietnam Vet: You know, with all the celebrations and congratulations being directed at us Vietnam vets, it is kind of like the parent who abandoned you at birth and now, 46 years later, wants to get connected and have a relationship. Just saying.

Army Vietnam Vet: I went from Pleiku, Vietnam, to Hanau, Germany, in February 1970. A year later, I took my discharge in Germany and traveled around Europe for three months before returning to the United States. It (mostly) wasn't in the United States that I got attitude. It was American draft dodgers whom I met in Europe. I won't try to share the narrative, but you can guess how unpleasant it was. A bunch of self-righteous draft dodgers who couldn't go home and wanted to blame me. I got a lot of questions about why I served in Vietnam from Europeans I met. However, I was able to reconcile with the issues in rational conversation, even with European anti-war activists. It was the American draft dodgers who were nasty, hateful, and spiteful.

Marine Vietnam Vet: Imagine yourself in a place where you are at the peak of your life. A place where every moment means something. Imagine yourself all alone there. Then, imagine the best people you will ever meet in your life join you. Now, imagine that you know that each second could be your last. That each thought could be your last, too. That any one of your friends could be dead at any moment. If you can imagine all of this you can feel for an instant what it was like for those of us who served in Vietnam.

Navy Vietnam Vet: Why is it that you think of that as another world, another time, and then something takes you back and you start all over again?

Army Vietnam Vet: When I first arrived in '69 there was still the will to win; when I left in October 1970 the attitude was, "Last one to leave, turn out the lights."

WAC Vietnam Vet: You know what saddens me? So many have died not knowing just how much we love them and appreciate their service.

Air Force Vietnam Vet: Vietnam has cost me two marriages. I'm on my third now. My ex-wives and my son and daughter just don't understand. I have woken up in cold sweats from dreams I had. I had dreams of nothing but blank faces. At times, I felt myself totally frozen and could not move. I could not fight back. I had one bad dream and drove my fist into the wall, barely missing a stud. My fist was sore for several days. My wife whom I am married to now has helped me a lot. I don't have them as much now.

Chapter 10 - Women at War

Women who served in Vietnam largely go unnoticed and unappreciated. In fact, 7,484 servicewomen served there. Of these, 6,250 were nurses and 1,234 made up the remainder. They were from all branches of service, officers and enlisted. They also served aboard navy ships and at bases in Thailand. They held jobs in clerical, logistics, transportation, intelligence, and personnel fields. Almost all were volunteers.

There were also many civilian women who served. They worked for various government agencies, civilian companies, and charitable organizations. Women in Vietnam paid their dues like any man who served there. Eight US servicewomen died in Vietnam, as well as 58 civilians, including 38 that died when an Air Force C-5A Galaxy crashed outside of Saigon in an attempt to evacuate Vietnamese children as the country was collapsing in 1975 (Operation Baby Lift). Many more were wounded, but I have not found any statistics on that.

Women who served in Vietnam were awarded many medals and citations, including Distinguished Service Medal, Legion of Merit, Air Medal, Bronze Star, Commendation Medal, Purple Heart, Commendation Letter, Vietnamese Award, and Unit Citation. But they were not recognized for their roles as their male counterparts in the service were. One woman, Karen Offutt, was recommended for the Soldier's Medal after she assisted in rescuing several Vietnamese adults and children from a burning building in Saigon. The Soldier's Medal was refused, and it was downgraded to a Certificate of Achievement for Heroic Action. The reason? The Army commander who reviewed the award refused it because, "We don't award the Soldier's Medal to women."

Karen was eventually awarded the Soldier's Medal on April 7, 2001, at Medard Park, Florida, in a surprise ceremony by Congressman Mike Bilirakis. She had no idea that this honor would finally be bestowed on her, 31 years later.

In 1967, the Women's Armed Services Integration Act was modified by Public Law 90-130. This law eliminated some restrictive regulations that applied to women. Previously, there was a two percent ceiling on the number of women who could serve in the military. Women had also been restricted from attaining the rank of General/Flag Officer. These restrictions were lifted. Women had a long way to go for equality in the military, but it was a start.

Unfortunately, many women veterans are still not accepted by some of their male counterparts. The truth is that female Vietnam War veterans suffer significant rates of posttraumatic stress disorder decades after the war. They were often subjected to sexual harassment and discrimination, in addition to being under attack by enemy activity such as rockets and mortars. Like their male counterparts, our Vietnam veteran servicewomen were also exposed to Agent Orange and other herbicides. Not only did this affect their health, but they passed on illnesses to their children long after coming home from Vietnam.

I started to cry. I always felt the VA was for the men.

Specialist 5th Class Glenda Storni-Graebe
WAC Detachment, USARV

I grew up on a ranch in a small California coastal town. We were all ranchers' kids. We went to school together, so I had no intention of marrying any of them, they were all like "brothers." I graduated from high school in 1965 and completed a semester of junior college. I drove a '53 Chevy, which burned a ton of oil. I had no support from my family, I had a job paying minimum wage and had very little money in the bank, so joining the military was very appealing. I intended to join the Navy because my boyfriend was in the Navy Reserve. At the testing station in Los Angeles, while waiting for the Navy recruiter, an Army recruiter started talking to me about all the great opportunities the Army had to offer, and before I knew it, I joined the Army. Besides, I would look better in green! It was December 1965, and I was 18 years old.

I went through basic training at Fort McLellan, Alabama, and AIT at Fort Leonard Wood, Missouri. My MOS was a Clerk Typist. After completing advanced training, I was stationed at Fort Myer, Virginia, and assigned to the Military District of Washington in Washington, DC. My best friend, who was the company clerk, informed me that she was going to Vietnam. I couldn't see her going without me.

When I learned my friend was going to Vietnam, I went to my colonel to ask for a transfer, but was denied. His reply was, "No! Women don't belong in Vietnam." My argument fell on deaf ears. One day, while he was out of the office, I typed up my own orders, signed his name, and ran it through the Pentagon. I got my orders and headed to California for my 30-day leave. My friend left for Vietnam in January 1967; I left for Vietnam in February 1967.

I enjoyed my 30-day leave with my family in Bakersfield, CA. But, my 30 days turned into 45 days, and I still had not received my travel orders. I finally called Fort Myer to locate my orders, and they were surprised to learn I was still in California. My orders were sent priority, and I was finally headed for Vietnam.

I flew to Vietnam on a commercial charter flight. I was the only female among 200+ G.I.s, a trip I will never forget. We arrived at Tan Son Nhut Air Force Base in the middle of the night. I stepped off that plane, saw soldiers with guns drawn, in full field dress lining both sides, and I was thinking to myself, "What in the world did I get myself into?" I could hear incoming mortar fire, and I could see the bomb blasts. I asked myself, "Am I going to die here?"

All the soldiers were being picked up by buses, but there was no one there for me, I was the only female in the airport. There was an E-8, who was more frustrated than I was. He didn't even know a WAC detachment existed and had no clue how he was going to get me there. While he was attempting to locate the WAC detachment, I had to sit there and wait. I was more frightened by creepy little beetle-toothed Vietnamese men than anything else. They had never seen a red-headed woman soldier and kept touching me, looking at me, and giggling.

Finally, around 4:00 p.m., he was able to locate someone from the WAC detachment to come and get me. The first field came to get me, and she was not happy to be out that early in the morning. The WAC detachment had not received a copy of my orders, so they weren't aware I was coming in. The only person who knew I was coming in was my friend who had arrived a couple of weeks before me. The detachment was located at Tan Son Nhut, until moving to USARV Headquarters in Long Binh, in June 1967.

The uniform of the day went from stockings, heels, and our summer uniform to fatigues, caps, and combat boots. We found it difficult to keep our fatigues starched and pressed and looking good. The living conditions were frightful and very dirty.

I extended my tour so I spent a total of 18 months in Vietnam. I was there for the Tet Offensive and experienced the real meaning of danger. I really didn't realize what kind of danger I was in. The night the ammo dump blew up was an eye opener. We had our helmets and all our field gear on. We didn't even have bunkers; all we had was sandbags to hide behind. We were watching all the firefights, explosions, and incoming. We had new recruits in-country, and I had taken my helmet off to give to a new girl. She was crying hysterically from fear. Some of us had been in-country for a long time and had become somewhat used to the noise, etc. I was down at the bottom barracks steps when I saw the explosion in the ammo dump. The concussion was huge! One of my Army sisters said, "That's okay, Storni. I'll protect you!" She came up behind me, and the concussion hit her back, and she rode me across the floor. The concussion went completely through our barracks, blew over all our standing lockers, and blew out the doors on both ends of the barracks. It was pretty scary that night. It was continual explosions for hours. We all spent the night sleeping behind the sandbags on the ground.

I decided to extend my tour for six months, and considered re-enlisting. I would have enjoyed a deployment to Europe; however, I was getting pressure from home that women needed to be married, and an Army career wasn't the thing to do.

I was given a 30-day leave for extending my tour, and flew home to be with my family over the holidays. When I landed, I was met by war protesters with signs and loudspeakers, cursing the government, military, and anyone who served. I was shocked, even though I had learned about the protesters.

I didn't have any problem getting home or even returning to the airport to return to Vietnam.

I haven't thought about this in a long time, but I couldn't wait to get back to Vietnam. When I was back there doing my job, I felt worth something. In Vietnam, I was home again, and I can kick myself because I missed Bob Hope and Martha Raye. They were at Long Binh, and Martha Raye was in the WAC detachment where I lived, and I missed the whole party. I saw pictures, and when our group gets together, they talk about how much fun it was, and I missed that because I felt the obligation to go home.

Back in Vietnam, I spent the remainder of my tour doing my job at USARV. In August 1968, in Vietnam, I was known as a "short-timer." I walked around with a "short-timer's" staff, strutting because I was going home in September. What an exciting thought, but I didn't realize what was waiting for me when I stepped off the plane. In September 1968, it was time to go home for good. My Army time was up, and I would be discharged when I got back in the States.

Coming Home

So, like the last time I flew on a commercial charter to Oakland, when we got to Oakland we were met by protesters! What was with this country? Protesting the war was one thing, but spitting, throwing garbage, yelling obscene words at us was another. All directed at us for doing our duty? Serving our country? I was shocked! If it weren't for the military fighting for their freedom, the protesters wouldn't be able to protest. What was this all about? This was the worst. I was more afraid of my own country than my tour in Vietnam. At least there we knew the enemy.

Arriving at Oakland, the plane was met by protesters throwing rotten tomatoes, spitting, holding awful signs calling us baby killers. It was such a traumatic experience to see my own country hating me for doing my duty. It hurt me so deeply. This is the hard part for me. I can talk about the bombs and the bullets, my job and experiences while in-country, but the pain I felt after returning has left a terrible scar, and I still have a difficult time discussing this experience. I love this country and was serving and doing my duty because I am a patriot.

We were told to step out of our uniforms for our own safety. In my own country! That was such a heartbreak to me because I was so proud of my uniform. I remember vividly stepping out of my uniform and changing into civilian clothes. I left everything behind. My uniform. The only thing I took with me was my Army-issued purse.

I went through processing (my Army time was up, and I was being discharged), and traveled home in my civvies, as instructed, only to find out that my family and friends didn't have a welcome home party for me. They didn't care about where I was or what I did, didn't want to know about what I had learned or seen . . . I felt so alone and confused. The disagreements about the war were one thing, but to hate a military person was another. I was proud of my time in the Army. I was proud to have been a part of a group of women first into a combat zone. We were history in the making, and this country didn't know about us or care about us being there and what we did.

Home

Little did I know this was the beginning of a downhill spiral of self-destruction. PTSD—what was that? There wasn't a place for women in the VA system; it was for the men. I didn't even know about the VA system, or how to get involved.

I married my high school sweetheart and had a beautiful daughter. Unfortunately, my marriage didn't last but five years. I didn't realize why I was having so much trouble in my relationship. After my divorce, I jumped into another relationship that lasted 12 years, 12 years of misery and alcohol abuse, continuing my path of self-destruction. And, I didn't know why. I became reclusive: didn't like crowds, didn't like noise, didn't like life. I came from an environment where I felt like I belonged and felt like I was contributing. I was proud of what I was doing. I came home to a family who didn't care. I had no direction and no support. I got into a relationship, and six months later, I said, "Oh man, what did I do this for?" I felt I had no purpose. I didn't have a career. Nothing. I had nothing. I became very angry and felt no self-worth: a failure.

In 1984, a friend helped me through the worst time in my life. I married him in 1985. He helped me survive the toughest times, and we are still married today. My husband would say, "My wife is a Vietnam veteran, and she's proud of it, and I'm proud of her." My reply to that was, "Shhh, don't say that, don't say that, people will get mad." And he'd say, "I don't care what they say, you are my wife, and I'm proud of your service in Vietnam." I had anxiety attacks. I'd be on my way to work and then feel like I was having a heart attack and didn't know why. So, I would be admitted to the hospital. I got to one point where I wouldn't go out into public because I was too frightened.

I was at my granddaughter's graduation, and we were going to the beauty shop to get our nails done, a Vietnamese shop. I'm sitting there, and they started talking in Vietnamese, and I kept getting madder, and madder, and more irritated. Upon leaving the shop, I jokingly told my husband, "I could just see the headlines tomorrow, a woman goes berserk and shoots up a Vietnamese nail shop." I heard myself say it; I meant it. Then I looked at him and I said, "I need help."

He insisted I go to the VA and get help. We went to the VA, checked in, and I sat down. I saw a doctor and I still kind of cried, because it's very emotional for me. The doctor looked at me, and she said, "Well, what has taken you so long to get here?" and I started to cry. I said, "I always felt the VA was for the men, they're the ones that deserve it." And she said, "You're a soldier, you deserve it too." But when we came home we were never treated like soldiers. At that point, she said, "We are going to get you help." They got me into a program, and I saw a psychologist twice a week. He got me to open up about things that I had held inside for all those years.

I was screened for PTSD and diagnosed. I was helped by therapy, discussion sessions, and a lot of patience, but was able to determine the cause of my troubles. All those years without help, without understanding, and sometimes thinking I was going crazy, I found out I wasn't angry at the Vietnamese; I was angry at this country, my family, and friends for the thankless behavior shown upon my coming home from war.

Now when I have an outburst of anger, an anxiety attack, nervousness about noise, or confusion, I understand the reason and can handle my reaction.

I am in contact with many of the women I served with in Vietnam. We don't talk much about Vietnam, we just have fun and enjoy our company as women with a special bond.

Everything in life is a learning experience. I have had some rough times in my life because of my Vietnam experience, but it's made me a strong person, a more understanding person, and I hope my story will help another fellow veteran.

Every time I see a Vietnam veteran I go up and shake his hand and welcome him home. I want them to know that I care. They are eager to talk, and I think now is the time for the Vietnam veteran and, unfortunately, a little late for a lot of them.

I served proudly and without regret.

Specialist 5th Class Karen Offutt
MACV Headquarters

I joined the Army in 1968. I was 17 when I graduated from high school. I went to Europe as an exchange student, and when I returned I enrolled in a school of nursing in Los Angeles. I was still 17 years old. Every time my friends and I passed an Army recruiting office, a couple of us decided that as soon as we graduated we were going to join the service and probably go to Vietnam. I guess we just wanted to do something with our medical training. At least, that is what I wanted to do.

I quit school the second semester. I was just too young. I didn't know anything about college. I was taking about 19 ½ units, plus working in the hospital. I went with my father to the unemployment office and there was a little recruiting trailer out in front, so I went in there and I joined the Army.

The recruiter in the trailer told me that being trained as a stenographer would be a good MOS. This was a really weird thing, because I didn't know how to type and I knew no shorthand. He promised I would go to Fort Benjamin Harrison, Indiana, for my training. I finished basic training and I was waiting for my orders for Fort Benjamin Harrison, but I received orders for Fort Benning, Georgia, to be trained as a communication specialist!

I knocked on my sergeant's door and, of course, she wasn't friendly. She said, "What do you want?" Then I said, "I have the wrong orders; there's been a mistake." She told me, "Write your congressman." So, I went back to my barracks and finally I had to go back and knock on her door again, which was like a walk to death, actually. She answered the door. "What do you want now?" and I said, "I'm sorry, but I don't know who my congressman is." Then she said, "Write to the effing president!" So, I went back and actually wrote President Johnson this long letter, about how my uncle had been in the service, my father couldn't get into service, my brother had asthma, my other brother was dead, and I was the one to serve.

A couple of weeks later, I was still waiting to leave for Georgia, and they called me down to the orderly room, and said, "You have a phone call from the Pentagon." I didn't know what the Pentagon was, so I said, "Okay," and I took it. This person said, "President Johnson has read your letter and he's very upset. You will get immediate orders to report to Benjamin Harrison, Indiana." From then on I was on the crap list, so I had to hide out in lockers and stuff. Then they sent a major over, and she put me in a little room and made me type all day until I could qualify for the 35 words a minute requirement. She said, "You better do good because we're going to be watching you. You caused a lot of problems."

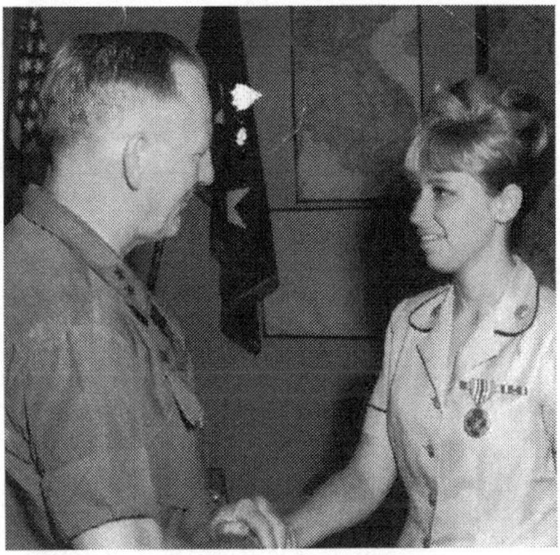

I went to Fort Benjamin Harrison, and I was so scared by what they told me that I ended up second in my class, with honors. I did 140 words a minute shorthand and they sent me to work at the Pentagon! That's how good I was. It was the fear factor. My MOS was 71C30.

I wasn't at the Pentagon long. I don't remember how many months, but I told them that I wanted to go to Vietnam. They said I was too young and I couldn't go. It was disturbing because there were a lot of protests going on and I started thinking, "How come only the guys are sent to die, to offer their lives? How come women aren't allowed to go?" I just kept bugging them, and finally they said: "There's going to be an opening for an executive stenographer and we don't want you to go, but we're going to let you."

So, I went to Vietnam in July 1969. We stopped in Hawaii, and some of the men who were going to Guam got off the plane. We then flew on to Bien Hoa. When we arrived, they told me to "Get on this bus." It had chicken wire over the windows. I asked, "What's that for?" When they told me it was to prevent hand grenades being thrown in, I looked back to see if I could get back on the plane!

Hot, red mud, and desolate. That was my first impression of Vietnam. I was supposed to be assigned to MACV in Saigon but was sent to the women's compound in Long Binh. It was run by older women. I was a kid, and I just hated it. I thought, "What the hell have I done?" One day I was walking off the post, and there was a big tall building. I don't know if it was the finance building or what. I went in there and said, "Somebody needs to help me get out of here. I was supposed to be with Major General Conroy at MACV in Saigon." So they called over there, and they sent a sergeant major the next day. The women who were in charge told him I couldn't go, because I had to be at least 20 to be in Saigon. So he yelled at me to get my stuff, get my gear. I didn't really have gear, but I went and got my little suitcase with a wool sweater (which I would never need) and he told me to get in the car.

I finally got to Saigon. In October, my roommate was supposed to get out and go home. My grandfather died at this time and I came back to the States for the funeral. I ended up with double pneumonia from sitting on the floor in Okinawa waiting for a cargo plane to take me home. When I got back, I was supposed to replace my roommate. She was General Creighton Abram's stenographer, but by the time I got back, that position had been filled. My new roommate was 37 years old. She scared the holy snot out of me. It was a funny experience, because I didn't feel safe from our men, I didn't feel safe from the women, and I didn't feel safe from the Vietcong. I just turned into my shell. We worked 12-15 hours a day, six-and-a-half days a week, and if they were going to do an air strike or if they had a meeting with generals, I had to do the dictation. I was just basically on call all the time. I was exhausted.

There are a lot of things that I don't remember. Many years later I kept having these nightmares, and I was having these visions and even auditory sensations about someone shooting at me. I thought, "This is ridiculous." I finally got on this website called In-country Women and we were discussing different things. I said, "This is so stupid; I must've seen a movie or something, and it's stuck in my head." These women said, "No, you were on the roof with us on New Year's Eve, and we were sniped at." I felt the bullets whistling by my ear and, yeah, that happened. They seemed unreal to me, you know? It is what it is.

When I was at MACV, I stayed in the Medford BEQ. It was an old hotel. We had two floors, and the men had two floors. It was so different than Long Binh because we didn't have anybody checking on us. I don't think anyone would have known if we died. We were basically on our own to do whatever. I was the youngest and the lowest ranking WAC there.

Coming Home

I received my orders to go home a couple of weeks before my scheduled DEROS. Actually, I was ticked off. My mother was going to have surgery for a hysterectomy, so they contacted me to come home a few weeks early. I didn't want to. I was mixed up and I didn't want to go home. I asked to do another tour, and they told me that to do it I would have to go back to the States for a full year first. I cried all the way home. I felt really guilty for leaving.

I left out of Tan Son Nhut. I don't remember the route home, but I ended up in Los Angeles somehow, and my parents were there. My mother had let her hair grow gray; she wasn't coloring it. She was going to have this surgery, and the first thing she said to me when they met me at the airport was, "Look what you've done to me."

I didn't have any issues with protesters, I think, because I was a woman in dress uniform and there were very few women over there. There were only 500 or so enlisted WACs who ever went to Vietnam. About 10,000 women served there, but over 90 percent were nurses. There were very few of us, and I think most of us are dead.

I don't remember seeing but a couple of old friends when I got back. I was really different then. I had been a song leader in high school, ran track, was Christmas Ball Princess, and all these things. I was really outgoing, but when I got back I was just in a shell. I was really sweet when I left. I remember somebody ran into my parents' car. I saw the damage, and the guy who hit them was there. I called him every name in the book and my parents were like, "Are you kidding me?" because I never cursed or anything. I had all this rage inside me for some reason, and I didn't know why. I went down to San Pedro, Fort McArthur, and while I was there, I married some guy I dated three-and-a-half weeks. That was the beginning of 16 years of misery.

Home

I was still in the Army, but my husband made me get out because he didn't like men looking at me. Then he and his mother decided that I should have a baby, so I had twins. I was married in August 1970, and the twins came in November 1971. One was born with cancer of the kidney. So, my life was pretty much cut out for me after that. I had a daughter later, in 1974.

All my children were born with cancer, epilepsy, bone deformities, or ADHD. And now my grandchildren have problems. My grandson has had diabetes since he was 9, my granddaughter has grand mal seizures and ADHD.

I testified in a congressional hearing, years ago, on the effects of Agent Orange and birth defects associated with the women who had served in Vietnam. I was one of the few women who testified there. The bill passed, but the guidelines for benefits are so stringent that our children practically have to have unicorn horns growing out of their heads to get any type of help or compensation. All the women at the hearing had children that were born with terrible birth defects: cancers, heart defects, and things like that, but I don't know of anyone whose children have received any compensation, unless it was for spina bifida.

While I was in Vietnam, on my half day off I went to an orphanage and helped with all the kids and babies. I was home when I heard Saigon fell. I just stood there and cried my heart out. All I kept thinking about were those kids and what happened to them.

I had all this rage inside me, but I just stuffed it down for a long time. I was dealing with the health problems my children were stricken with and my husband had received orders for Vietnam. He had them changed. He wouldn't allow me to talk about Vietnam, or look at my pictures, or any of the things I brought home. He threw a lot of my stuff away, which really hurt me later on.

I called the VA about Agent Orange and they laughed at me and told me to call Washington, DC, if I was concerned about it. I finally went to a meeting where they had a spraying map, and I said: "I don't think I was sprayed. They were spraying for mosquitos and bugs, but I don't think it has anything to do with my children's cancer and epilepsy." They asked me when and where was I there, and I told them. They showed me on the spraying map that I was at the core of it, the year that they sprayed the heaviest throughout the whole war.

I started having panic attacks. I actually thought I had a brain tumor. I didn't understand what was going on with me. The first attack was so horrendous, it felt like three or four gunmen coming in and pulling guns on my family. That's the kind of panic I felt. It was sheer horror. I went through a lot of tests, and they said, "There's nothing wrong with you." There has to be, with the nightmares and the anxiety. I was always being pursued by Vietcong, or I was looking for guns. I didn't have a gun in Vietnam. I had a sword and a billy club, and hoped for the best. I didn't know anything was wrong.

I didn't start getting help until the 90s, maybe late 80s. What are you going to do for PTSD? Seriously, I still go to therapy every other week, sometimes weekly. There's no cure. It only helps in the fact that I've lived alone for over 16 years, and I don't really talk to anybody. I mean, I don't have any human companionship. Sometimes it helps just to talk to somebody. With the VA, they transfer you out, or your doctor quits. Then you have to go through the intake process again. "When were you there? What did you do? What did you experience? Blah, blah, blah." That in itself is painful, and I think it does more harm than good.

I have been to the Wall and that was very hard for me. I know a number of people on the Wall and I know a lot more who should be on it. I helped form VVA Chapter 441 back in the 80s. Most of our guys are gone. Some of them killed themselves. There are a lot of people who belong on the Wall who aren't on it. It's all sad. Part of my rage is that these wars are meant as money makers and we're expendable.

Sometimes, I will be asked to give a talk on the war. I say, "Yeah, but if I do that, I will only tell you the way I feel about war." I don't care who you are, you don't come home the same. My experiences are nothing compared to what other people did and saw, and the guilt that they harbor from killing people and watching their buddies blown apart. How do you get over stuff like that? You drink, do drugs, kill yourself, or you isolate. I isolate.

If you saw me driving on the street and heard what I say! Other women don't act like that. Maybe, but it's hard if you live alone . . . I just have animals. I have a lot of animals—I help rescue animals—but if you live alone, and you don't have anybody to laugh at a movie with or talk about other things, your mind just keeps going all the time. That's why I don't sleep at night. I don't sleep, period. If I sleep at night, it's like for an hour, then up again. I'm still awake when daylight comes. I can sleep when it's daylight. I could sleep maybe five hours, but I can't do that either. I have so much to take care of here by myself. I only feel safe when it's daylight. That's my life. I have a few girlfriends and they don't live like that. You're always on alert. I remember when I got back and I was working at Fort McArthur, I had my dress uniform on, and I went to this store and somebody dropped a can. I hit the floor and started crying, and I was totally humiliated. I mean, absolutely humiliated. Or, being in a dressing room, trying on clothes, when somebody slams the door, and I scream. And that stuff still happens.

I used to wonder why, when we went to a restaurant, I had to have my back against the wall and as far back as I could get so I could see the whole restaurant or be near an exit. Then when I got divorced, I was working as an RN, but in the middle of the night I would just take off and drive as far away as I could. I would park somewhere, look at the stars and try to get my head together. Then I'd come back home. When

I got with this group, the VVA, I was talking about things like this and said, "This is not normal, what I do." They're all cracking up, because they all do the same things. That was the first time I realized that other people did those things.

Still, when I go out to a restaurant with my daughter, she knows to get a table where I get to see everybody. Those are things I didn't read about. I didn't talk to anybody about those habits, that is just typical of things I did, that I knew were weird. I've been criticized by some Vietnam vets; "You know, you had clean sheets over there, so what are you bitching about?" They want to kill Jane Fonda. Put your hatred somewhere else. I mean, you know, she was young and stupid, and did whatever they told her to do, and half the stuff they said happened probably didn't happen anyway. I'm not going to sit here hating some old lady that's 80 years old for something she did when she was 20.

My second marriage was to a guy who was in VVA Chapter 441. The real reason I married him was because we had so much in common, being Vietnam vets. We had rap sessions at my home, and our Vietnam veteran friends would come. My husband was shot twice and he was an LRRP. He would cry at these sessions and talk about the first time he had to slit a child's throat to keep her quiet, and all this stuff. He lost his job immediately after he married me, and I was still working to support my children. I was paying his child support from his other two marriages. He was in bed uncovered one morning, because he's sleeping while I'm getting up for work. He had said he was shot in the butt one time. So, I'm looking at his butt for the first time, and then I woke him up and said, "Hey, John, weren't you shot in the butt?" He said, "Yeah," and I said, "Well, how come there's no scar?" He said, "They did plastic surgery on me." And I said, "BS! My tooth fell out while I was in the Army, and they didn't do squat." Then I realized everything he said was a lie. I gave him until midnight to tell me the truth or leave. He finally admitted it. His ex-wives and his parents never knew that he was not in Vietnam. I've met so many fakes since then that it's not even funny, even women.

I don't understand these fakers. When I do give my talks, I say, "Who in the hell would want to be a Vietnam vet?" Seriously? It's a liability in every aspect. The only thing it did was make me a tough cookie, but I really would rather be the sweet person I used to be. I don't know how to get that back. It can't come back, I think. Probably only with my family and the animals that I rescue am I softer.

It's funny, because I tried to go to PTSD groups and a few were okay, but for the most part the men were hateful, crass, and condescending. If you were a woman and you went to Vietnam, it lessens their manhood or something. I don't get that. So, I can't go to any groups anymore, even though I really want to. I like sharing experiences with people and seeing how they handled things. When you're a woman, it is hard. There aren't any women here like that. Who are you going to talk to?

I guess they just don't want to hear from a woman that she went through anything. That's the hardest part of being a woman Vietnam vet: you don't fit in, and I think the women who did survive better than I have were older, lifers, or they stayed in the service for a long time after they got back. They have that camaraderie and that routine that they stuck to, and I think it helped them.

Vietnam changed the way I felt about a lot of things. First you become more religious, because you think you're going to die every day. Then, when you see what I saw, at the orphanage, and all the coffins lined up at Tan Son Nhut, and going to the 3rd Field Hospital, seeing the guys with their faces blown off, arms and legs gone, it makes you question a lot of things. I've never turned my back totally on God, but I did turn my back on religion. I'm not part of any organized religion anymore, because I see people who are too hypocritical and judgmental. I think I'm more spiritual, as much as a cussing old lady can be. I talk to God every day. I try to be good to homeless people and homeless animals, and I do what I can do to keep in contact with my God. It definitely makes you look at things differently. Look at today, it's still the same thing—bombing children—and it's sad.

I think I get the most tired of people saying that Saigon was a safe place. Look at the Tet Offensive. What happened in Cholon in Saigon?

Cholon was still a wreck when I got there, and look what happened when they took over Saigon again. We had claymores planted. We had Jeeps blow up. We had crashes, we had snipers, rockets and mortars. In Long Binh, my first night, they pounded the hell out of us. I just hate it when people say, "Oh, you were in a safe place."

I am proud that I served. It's weird when people say, "Thank you for serving your country." I don't know that I served my country. I served some bureaucrats in my country, but nobody was crossing our borders. To stop the spread of communism is what I was hoping to do, but that's not what that war was about. I think that war is a testosterone- and money-based idea or ideal. The children of the rich aren't going to see war. They're not going and they won't see that it changes you forever. Whether you want to or not, whether you say you're okay or not, you're not okay. You won't be okay. I don't care in what capacity you serve, if you spend time in a war zone, your life is going to be changed forever. So, if you go to war, it better be a worthwhile endeavor, like to protect our country.

I felt restless, I needed to move. I had bouts of depression.

1st Lieutenant Grace Oberholtzer

93rd Evacuation Hospital

I served in the US Army Nurse Corps, which was a separate organization at that time. Both of my parents served in World War II. My mother was also a nurse.

I graduated from nursing school in 1968. The nurses I went to school with were making $7,000 a year with a two-week vacation. That wasn't for me. I liked the idea of traveling and seeing the world. An Army recruiter came in and gave his talk. Of course, he didn't say we would go to Southeast Asia.

I joined the Army in January 1969, and entered as a registered nurse. Everybody that was a registered nurse went in as a 2nd Lieutenant. Our basic training was at Fort Sam Houston, Texas. It was totally different than that for men and WAC's. We all had our own rooms. We learned about the Army, learned how to salute, how to walk, and we learned how to march. I never marched again after basic training.

My father was very proud of me and he wanted me to go to the service. The day before I was leaving to go into the Army, I found him standing against the wall and he said, "The pain won't go away when I'm lying down." Oh God, here am I, my mother and I are fighting to get him to the hospital, and when we finally got him there we found out he had had a massive heart attack! My mother said, "Now, you go, he's out of ICU. Go into the service." So, I reported a couple of days late. I had been at Fort Sam for two weeks and I hadn't heard anything from any of my friends. Finally, I called my mother. She let something slip and said something about my father's pacemaker. I hung up the phone, turned right around and called her back. I asked her, "What pacemaker?" I then found out why nobody was talking to me. My father had coded 44 times and she had banned all my friends from letting me know.

So, anyway, I received orders for Vietnam. I left out of Oakland, California. I was very lucky. There were a few people with me on that flight that I would be stationed with. Some of the guys took good care of us nurse lieutenants. We stopped in Hawaii, where we had Mai Tais while waiting for the plane to refuel. We then stopped in Guam, and one of the people on the flight knew someone there who came to the airport and gave us all sake. We were drinking at every stop we had. We played poker on the way over on the plane with the new command sergeant major.

We arrived at the 90th Replacement Battalion and I was so tired. I told my friend, "I'm exhausted, I'm going to go to sleep." So, I walked into the shower, and there was a cockroach as big as my fist! I called to my friend who was going to the office to check in, "Wait for me. I'm coming with you." So, I went to the office at six o'clock. I do not remember coming back to the barracks. I was so exhausted I slept through a mortar attack, the only mortar attack I was ever in!

The next day we went over to the 90th Replacement Battalion and were assigned to the 93rd Evacuation Hospital in Long Binh. I was assigned to the intensive care unit. The unit had 36 beds: 24 intensive care and 12 recovery room beds.

I met my ex-husband while I was there. He was a Dustoff pilot with the 45th Medical Company, Air Ambulance. They shared our facilities. When I arrived at the 93rd, people were very standoffish. They did not want to include you in anything. They were short-timers and they had already made their friendships. It was difficult to break into that scene. After I got there, I made sure everyone felt welcome. I didn't want anybody else to feel like they were an outsider.

The patients were very, very easy to take care of. A lot of them were very, very sick, and you tried your best not to show them how sick they were. I knew one of the first patients I had was going to die. He was in the artillery, and operated a howitzer. I figured, after this patient I can deal with anything. The howitzer misfired. He was a double amputee from the thigh down, had abdominal wounds, and was burned over 60 percent of the rest of his body. He was awake the whole time. We got him to Japan, and it took me a month to ask somebody what happened. He lived over there for two weeks before the infections took over. After that, I never asked. It wasn't that I didn't want to know. The only way to do my job successfully was to say that my patients made it to Japan, and then it became somebody else's job. That was it. My job was to get them to Japan.

Coming Home

I was married after my tour. I should never have gotten married. We were the best of friends in 'Nam. Getting married was the biggest mistake, because we would still be friends today if we had not married. We stayed married for two years. I know he had really bad PTSD, and he was having a very, very difficult time fitting in. He couldn't even talk to me about it. My husband and I have reconnected. He is happily married and has four children. We have friends in common, and have now established a platonic friendship.

My girlfriends were great. They would take me out drinking, taking me here and there. Some of them were married and didn't go out with us as much. I had 39 classmates that graduated with me from nursing school. Before I left for 'Nam on December 6, everybody put up a Christmas tree to wish me well. Every single one of my classmates put up their Christmas trees early so I could have Christmas with them all. When I came back, they were still willing to do all that stuff with me because they were nurses, too, so they knew what I had gone through. Some I stayed closer to than others. Everyone had gone in different directions, but they were there for me. I could stay at anyone's house if I wanted to; however, I still didn't fit in.

Every once in a while I would say something to someone about Vietnam. You know, funny stories, but they would say, "We don't think you did anything over there. You didn't do this. You didn't do that." I would look at them and say, "You know, I really feel sorry for you and really feel that you're pulling me down, and you're making light of me, and I don't know if it's because you can't deal with the fact that there was a war and that I served."

A lot of my nursing friends did not want to hear anything about the war, not even the funny things. There were some I could talk to, but there were a lot that I couldn't. I am not sure why: perhaps guilt? I had one friend that I would never talk to about Vietnam because her brother was killed there while we were in nursing school. I can still remember her screams in the dorm. Yet she was one of the first to welcome me home and be there for me.

I felt restless, I needed to move. I had bouts of depression. After my divorce, I moved to Vermont. I got my head together there. I read a book after I got married the second time and my kids were small. It said this was a symptom for Vietnam nurses, they felt they had to run and they were running. They did not know what they were running to or from. Restlessness. You're running to find something. I could recognize myself in this book. Did I ever figure out what I was running to or from? Yes and no. I was running because I did not want to be around people that I loved. I felt that I would bring them down with me because I would have periods of depression.

When I returned from Vietnam, I planned to move to Denver because I needed to move on. I had classmates living there. So, I went there, got a job, and planned my move from home. The day before I was going to leave, my father had a massive heart attack and died. I looked at my mother and said, "I won't go." She said, "You're going. I'm 50 years of age and I can take care of myself. You're not living here. Go." So, I went to Denver. My best friend said to me; "If you wait a year and go to California, I'll go with you." I said, "No, I need to get out of here."

After eight months in Denver I moved back to Pennsylvania to be close to my family and friends. It was there I met my second husband. We were married in Pennsylvania and moved to Chicago. We had a great marriage, almost 24 years. He gave me two beautiful children. He was in Vietnam in 1966. He was an Army lieutenant, a Ranger. The strange thing is that one of his classmates was Patton's son, George. He got to meet Mrs. Patton. He and all of George's friends were never assigned to combat units. My husband was assigned as the XO of the 90th Replacement Battalion. He developed prostate cancer and then got pancreatic cancer. Pancreatic cancer is not recognized as an illness from Agent Orange exposure. My husband died when he was 65. My best friend from high school died when he was 65, and another classmate of mine from high school died at 65 from pancreatic cancer, also. All three served in Vietnam at the same time.

I have colon cancer. I was very lucky I was diagnosed at stage one. However, the cancer's not recognized by the VA as a result of Agent Orange. The last time I went to the VA they said to me, "You're not entitled to anything." I said, "Thank you very much," got up, and walked out. I thought, "Screw you, lady, you don't know me."

Home

I don't mind that I served over there. I have blocked most of the patients out. When I go to the Wall, I don't recognize one name there, though there are patients on there that I took care of.

I will relate a story. There was a gentleman that I took care of, and most people today never heard this story. We were a burn unit, and we would wrap the patients with dressings, and they looked like they were mummified by the time we were done. We wrapped their abdomens, arms, legs, even their faces if they had facial burns, because they have to go all the way with that dressing to Tokyo.

We had a young gentleman—I shouldn't say young, he was a college grad. and had been an English major. He had stepped on a Punji stick. It infected him, and because he was so intelligent, the idiot doctor did not take off the leg. They should have taken off the leg. The infections became worse, and we knew he was not going to make it.

At night, when they were evac-ing patients, this young man started talking about something from Egypt, and about mummifying the bodies, and taking them out in the dead of the night. My corpsman started making fun of what he was saying. And I turned around and put him to the wall. I said, "This is his way of seeing it, he is at his time and he's feeling for these people. I don't want to hear one more word about this. I want you to go and apologize to him. You're not going do this to him. He is a human being and you will give him the dignity that he deserves. This is the way he's dealing with it".

After that I thought, and I said, "You know he's going to die. We all know he's going die. He's going to wait to die until I am alone with him." I was working days at that time, and my sergeant and I looked at each other, and I said, "Don't leave him alone. There will always be a nurse with him." It was at night and at 12 o'clock it was dinner time. It was the first time he was alone for a week. As soon as the nurse left the unit he died. I was told, "Lieutenant, he died a moment ago here alone." He died with dignity and chose his time.

That's the purpose I felt that we were there for. You know, whenever you touched somebody, you were there for a reason. The only anger I have is the illnesses Agent Orange left us with.

Chapter 11 - The Wall

"They shall not grow old, as we that are left grow old:
Age shall not weary them, nor the years condemn.
At the going down of the sun and in the morning,
We will remember them."

Laurence Binyon, 1914

In 1979 the Vietnam Veterans Memorial Fund was started as a non-profit organization to construct a memorial honoring the veterans of the Vietnam War. Jan Scruggs, a wounded Vietnam veteran, was the driving force behind the memorial. After raising over 8 million dollars, Congress authorized the building of the memorial on a two-acre site near the Lincoln Memorial.

A design competition was held, and there were 2,573 entries competing for a $50,000 prize for the winning design. The selected design was submitted by a young artist named Maya Lin. She was 21 at the time. The design was shrouded in controversy. There were many who did not like the design, and initially the Secretary of the Interior refused to issue a building permit. It was called an ugly gash in the ground by some.

The Wall was completed in 1982 and dedicated on November 13 of that year. The Three Soldiers statue was added and unveiled in 1984. The Vietnam Women's Memorial was dedicated on Veterans Day: November 11, 1993.

The Vietnam Veterans Memorial (the Wall) is now one of the most visited memorials in Washington, DC. There were over 5.6 million visitors in 2015. At the dedication in 1982 there were 57,939 names engraved. As of Memorial Day, 2015, there were 58,307 names. The additional names are of those who were wounded during the war and later died of their wounds.

The Wall is a healing place for Vietnam veterans, a place to put some of their demons to rest. This "ugly gash in the ground" is considered a place of supreme reverence by ’Nam vets. It is a place of solitude, strength, and sadness, a place where we can speak to those who have gone before us and will forever be young, a place to say goodbye and finally mourn the loss of friends, to mourn them like we couldn’t back in our young days, back in Vietnam.

Army Vietnam Vet: I went to the Wall. I can never go back; ain't gonna go through that hurt and anguish again. Just can't.

Veteran Widow: It is overwhelming to think each name you're looking at was a real live person at one time, someone's husband, son, brother, nephew, niece, and friend.

Air Force Vietnam Vet: I went to the Wall, and as I was standing there with tears streaming down my face, a fellow vet came over and put an arm over my shoulder and we both just stood there. I couldn't tell you what he looked like, we never said a word. In a few minutes we shook hands and went our separate ways! I will never forget that moment in time!

Navy Vietnam Vet: I went to the Wall for the first time on October 17, 2015. I don't know exactly how long I spent there, probably close to 45 minutes, and then I waited until there were no other people at the Three Brothers statue. I sat on one of the benches there and just watched the people come, giggling and laughing, taking their pictures, having an enjoyable Saturday morning, and then I watched them walk away, clueless. Don't mean nothin'…

Army Vietnam Vet: I have been to the Wall. There are the names of the men I knew who were killed. I knew them so those names mean a lot to me and the others that knew them, especially their families.

Marine Vietnam Vet: I saw the Traveling Wall in California. I spoke with the guy who got the project done and thanked him for his efforts to honor our heroes. As a Marine air winger I just did my little thing, taking care of Tacan test sets in-country, so I sure was no hero. It was sure tough to walk up to the Wall, even the traveling one, 'cause it represents so many who never came home.

Army Vietnam Vet: Very emotional. Been there a few times. Brings back too many memories.

Army Vietnam Vet: I have been there once. It was 1:00 p.m. and as I stood there I noticed that I was the only person standing in front of this big black granite wall. I laid down a camo shirt and sat on the ground under the name I was looking for. I waited for a park ranger to help me get a name rubbing. No one came. I finally left and made my way to the Lincoln Memorial, where there was a booth selling blank paper and pencils. I went back to the Wall and did my pencil rub, and again I noticed I was the only one around. There were hundreds of other people and park rangers, taking pictures, etc., at the other memorials. I just felt so alone and sad. I left and have never gone back. I guess people only show up on those special days when there are news cameras or some politician wants to be noticed for some reason or other. Now I'd better stop, I'm starting to get mad.

Army Vietnam Vet: The first time I visited the Wall it was a cold and rainy day. It was a very trying thing for me, it brought back so many memories of the brothers who are engraved on the Wall. So many were just young men just starting in their lives. Such a terrible waste. God bless them all.

Widow of a Vietnam Vet: Other than the rows and rows of white crosses and Stars of David in WW II cemeteries in Europe and the one in Hawaii, this has to be the saddest place on Earth.

Air Force Vietnam Vet: Been there many times. It is the most solemn place I've ever been. I hope to make it back one more time.

Army Vietnam Vet: I went to the Wall with about 75 of my guys from 'Nam, in November 2009, and I was okay with it. It shakes you somewhat, but you feel better the rest of your life. Down that wall all the way towards the Washington Monument is the name of one of my classmates from high school, who was killed in 1966.

Army Vietnam Vet: Been to the Wall a couple of times while I was at Fort Monmouth. It's a very sobering experience for this old 'Nam vet. God bless all the brothers and sisters who are on the Wall. Just saying thanks is never enough for the hell we all went through.

Navy Vietnam Vet: I felt like this for almost 34 years, then I went, and, yes, I did, but it didn't feel like I was out of control . . . just terribly, awfully sad.

Widow of an Army Vietnam Vet: I went with my husband: we both cried.

Navy Vietnam Vet: Been there once, don't care to return—too emotional.

Marine Vietnam Vet: Never been there, brings back too many memories.

Army Vietnam Vet: Two years ago, my wife and I visited Fort Benning and unexpectedly found the Traveling Wall at the Infantry Museum there. As I approached the Wall I began to tremble. It was a very emotional and powerful experience. However, I did not find any names as my vision was clouded that day. On April 21 I will be on the Honor Flight to DC with a group of Vietnam veterans. I will find some names and leave a remembrance. I am already at ease just knowing once again I will be with my brothers.

Sister of an Army KIA: I've visited many times, from before daylight until after sunset. My brother's name draws me to it like a magnet, and when I touch that wall the strength of all those whose names are etched on it seems to wrap around me and my soul is soothed.

Army Vietnam Vet: When I heard they were open to ideas about building a Vietnam memorial, my thought was of a giant, five-ton, cast iron "V." And let it just stand there in the elements and rust away. And until the very last of it was gone it would be a reminder to America that we should never fight another war in that fashion again. We should never fight another war unless it's a declared war by Congress. We should never fight another war unless the entire nation is committed to it. We should never fight another war where the rich kids are exempt from it. We are either all in or we are not in at all. That's my thoughts. Bitterness on my part, I suppose. I've never been to the Wall. I've been to the Touring Wall. It'll put you on your knees. I don't know if I could go to the Wall on the Mall. Bless you that have. It is solemn, beautiful, and respectful, much better than my idea.

Chapter 12 - Australians

Australian military involvement in Vietnam began with 30 advisors in 1962. In 1964, the Royal Australian Air Force sent a squadron of Caribou transport aircraft and supporting troops to Vietnam. By the end of 1964 Australian troop strength had grown to 200. Australia instituted conscription at the end of 1964 to fulfill the growing needs in Vietnam.

In 1965 the Australian government decided to back American efforts to stop the communists from taking over South Vietnam. This decision set the stage for a much larger commitment, including combat troops. Infantry forces began to be sent to Vietnam along with their support personnel. The Royal Australian Air Force supplied transport aircraft, bombers, and helicopters to supplement their ground forces in operations against Viet Cong and North Vietnamese troops.

As their military presence in Vietnam increased so did opposition to the war, much like the situation in the United States. There were demonstrations and protests against Australian involvement in the Vietnam War. As the anti-war sentiment grew, the Australian Diggers were subject to the same scorn and mistreatment that returning American veterans were.

By late 1972 all Australian forces were withdrawn. During the war, they suffered 521 deaths and 3,129 wounded. Many Australian troops suffered from the effects of PTSD and Agent Orange related illnesses. In many cases, the returning veterans were shunned by World War II veterans and were excluded from joining the Returned and Services League (like our VFW). Many Vietnam veterans were also excluded from marching in the annual Anzac Day parades during the 1970s.

Agent Orange illnesses are common to Australian veterans, including their blue water Navy personnel. Australian ships were exposed to the herbicides due to aerial drift as a result of direct spraying. In 2002, a study by the Australian Department of Veterans Affairs suggested that the water distillation process used aboard Australian Navy ships resulted in increased concentrations in the potable water made aboard ship. In reality, the mortality due to Agent Orange of Royal Australian Navy personnel was higher than that of veterans who served on land. The Australian Department of Veterans Affairs is now providing disability compensation to its Vietnam naval veterans for medical conditions related to herbicide exposure.

Warrant Officer 2nd Class Terry Sullivan
Royal Australian Army, Retired

I was born in Great Britain. I served in the Royal Marine Commandos-Special Services from 1962 to 1965. I was serving with the Special Forces in Borneo when my parents and sister emigrated to Australia in 1965. I joined them in Australia, and joined the Australian Army. I served in the Australian Army from 1965 until my retirement in 1986. I then served in the Australian Army Reserves from 1986 to 1999, and the Australian Army Cadets from 1991 to 2006.

I chose a military career to continue my family tradition. During World War I both grandfathers served, one in the Royal Navy, the other in the Army. My father served in the Army in World War II. Now my son serves in the Australian Army. He has served in both East Timor and Afghanistan.

I volunteered for Vietnam and was assigned to the 5th Battalion Royal Australian Infantry, where I served from 1966 to 1967. I was in an infantry section as an M-60 machine gun carrier.

Coming Home

I left Vietnam in August 1967. We came home on the HMS Sydney, an aircraft troop carrier. I came home with my whole unit. The trip home on the aircraft carrier was pleasant. We had time to ourselves, but there was tension in the air knowing what happened to the 1 Battalion Infantry, who were the first troops in Australia to come home from Vietnam in 1966. They were harassed during their march through the city.

We arrived in Sydney on August 30, 1967, and it was a nightmare. We marched through Sydney and were booed. Someone threw red paint at a few soldiers ahead of me. We were called baby killers and other rude comments.

My emotional state then was hard to explain. As a young man, you feel proud of what you thought you had accomplished for your (my new) country. I was happy to be going back to Australia, having only spent five months there before going to Vietnam, and previously 12 months in Borneo with Special Forces.

People just didn't understand and would not listen. We were shunned by people at gatherings. The RSL (Returned and Services League, like the VFW) didn't want to know us. We were constantly harassed by university students. No one had time for the soldiers returning home from Vietnam. Much different than going back to England from Borneo. There was no welcome in England; just quietly back into the country, as you would expect.

Home

It was very hard to become connected to my family on returning. I hadn't spent much time with them, having come directly from the Royal Marine Commandos-Special Forces straight into the Australian Army. I did infantry and jungle warfare training, then was sent to Vietnam after only five months in Australia.

My problems started when I tried to merge with civilians, but found that I could only mix with soldiers. These problems started (though I didn't realize at the time) almost immediately. I started to drink very heavily and all I wanted to do was play hard sport.

On returning from being out on exercise with the battalion for several weeks, I wouldn't go home straight away, but drink for a couple of hours first. Needless to say, my first marriage broke up. I was very argumentative, self-centered, and could only see what I wanted to. I found that I was terrible company and didn't like mixing with too many people. I sought help after my soldier friends told me I had a big problem. They had already sought help with specialists.

My help started in 1970 and is still ongoing. I have had prostate cancer and had it removed. Agent Orange was sprayed during my tour near our area. We also passed through many areas that had been sprayed. But, to this day, I am not sure about the effects of Agent Orange. Doctors seem to shy away from this type of question.

Looking back at the war I feel that it was all a waste of time, having lost several friends, soldiers that I had gone through training with.

Experience: it proved to me that the only person you can really trust is the man in front of and the man behind you when you're out on patrol in the jungle. I don't give trust that easily.

I kept seeing the men with missing limbs and the coffins and thought to myself, 'What was the purpose?"

Corporal Peter Goldsmith
Detachment 11 Movement Control Group

I had to register for National Service in 1966. In the beginning of August 1966, I came home from work to find my mother sitting in a lounge chair, and when I walked in she started crying her eyes out, and I said, "What's the matter?" She said, "You've been accepted to go into the Army." It was heart-wrenching. That's how I found out I had been called up. My dad was pretty upset as well because, as you're aware, Vietnam was the first war that really came back home through television, and television was not that old in Australia at that time. So, they had seen the war on TV and knew what was happening there.

When I was called up in 1966, it was relatively early in the grand scope of the Vietnam War. At that time, the feelings in Australia about the Vietnam War were sort of half and half. Half were still very much in favor of being there, because we were told it was about stopping the communists coming down towards Australia. The other half were starting to oppose conscription and being in Vietnam. They were saying, "We really don't want to send our young people over there to end up coming home in a coffin." So, opinion on Vietnam was mixed, but it was starting to turn.

I was called up into the Army on September 28, 1966, for a period of two years. My recruit training was ten weeks long, and at that time you could register for three choices of different jobs that you wanted to do. I made my choices then and sat down in front of the gentleman who gave the corps assignments. He told me I was going to the engineers, which wasn't one of the corps that I asked to go to.

So, I had to quickly find out what engineers did. The training was ten weeks long and that involved learning how to build bridges—Bailey bridges in particular—pontoon bridges, and instruction on landmines, booby traps, how to use claymore mines, and how to use improvised explosive devices. After training, I was assigned to the Detachment 11 Movement Control Group (Det 11 MC Gp), which was one of a group of three units consisting of the Det 11 MC Gp, Detachment 30 Terminal Squadron (Det 30 Term Sqn) and 32 Small Ships Squadron. These three units were responsible for moving personnel and equipment in and out of Vietnam from Australia.

I pretty much knew I was going to wind up in Vietnam at some point because of the small number of call-ups. We were told that eight out of ten of us would end up within Vietnam somewhere, so it was no surprise when I received my orders to go.

I married my first wife on January 6, 1968, and went to Vietnam on April Fools' Day 1968. I actually felt happy about going to Vietnam, because I knew I was going to contribute after all the training I had undertaken. I was looking forward to that, but I was sad that, from my parents' point of view, if something happened to me while I was over there that would be traumatic to them, as well as to my wife.

I flew into Saigon on a Qantas flight from Australia as a replacement. From there I caught a Caribou down to 1st Australian Logistical Support Group (1ALSG) base at Vung Tau, where I would spend most of my tour. The guy I was replacing was there for another week training me, then he flew out and I took over his job. The normal tour was 12 months, but I was set to be discharged in September, so my tour was shorter.

Our Hercules planes flew into Vung Tau with replacements and some of our supplies. Our responsibilities were the coordination and control of the movement of passengers and freight on RAAF flights, on weekly Qantas flights between Vung Tau, Nui Dat and Saigon, of air movement of personnel on R&R flights from South Vietnam, passenger and freight shipping, coordination of tasks for Landing Ships Medium (such as Clive

Steel), and the coordination and operation of the Military Forwarding Office.

Once the planes had been unloaded, two of the worst activities we had was to load on the wounded who were flying back to Australia for further medical treatment, and the coffins of those who passed.

We did come into contact with the Americans at the base there, and a number were surprised that Australians were involved. But, we just sat down and we had a good chat and, at the end of the day, we got on well together. The Americans were amazed at the training we had had prior to going to Vietnam. We had ten weeks of recruit training, ten weeks of corps training, and two weeks of jungle training in Queensland. This was all part of the process before we left Australia. The training was fantastic, and we also received a pocketbook before we went over that the Australian government produced, giving us insight into the history and geography of Vietnam, Vietnamese culture, and some of the language.

One problem we had when I was at Vung Tau was that the wharfies (people who work on the wharfs in Australia) went on strike, so we didn't receive any mail or any food supplies until the strike ended. That was pretty hard to take, knowing that our people back home decided not to support us by going on strike at the wharfs.

Coming Home

In September 1966, I received my orders to go home, and spent the next week or two processing my paperwork and training my replacement who had just come over. After I trained him for a week, it was time to leave. I flew on a Caribou from Vung Tau to Saigon, and when we landed the two pilots came out and inspected the plane. As we were hopping off one pilot said to the other, "They weren't there when we left." We said, "What wasn't there?" He pointed up, and when we looked up at the plane there were four holes. Someone had shot at us either leaving Vung Tau or landing in Saigon. I was very happy to get off that plane.

After arriving in Saigon, we boarded a Qantas flight for the trip home. The only thing that killed it for us on the way home was we only had one can of beer each because the guys coming over drank all of the supplies, and it was a ten-hour flight from Saigon to Sydney! We landed late at night, maybe about eleven o'clock. We had to go through customs, and they were absolutely fantastic. They pushed as through as quickly as possible, then we went to the arrival center, and in my case my mom, dad, and my then-wife Judy were all there to greet me. I was in my dress uniform and the reception was good, but when we walked out of the airport I was spat on. There was a group of demonstrators—some called Save Our Sons—out front and they called us child rapists, child killers, etc. By 1968 it really started to turn. That was a real shock to us.

I still had a couple weeks left in the Army, so I had to report to my unit the next day. My family lived in Sydney, so I hopped on public transport to check in at my unit. I was in uniform and again I was abused and called the same names: child killer, child rapist, etc. "Why did you go?" "Why didn't you stay home?" and, "We don't like what you're doing over there," etc. Right there on the bus. I didn't say anything, to be honest. Actually, I was what we call gobsmacked: I couldn't really believe the reception. I was jet-lagged, exhausted, and shocked.

I did not have to stay at the base and was allowed to go home every day. I had to put up with this abuse every day for a couple weeks. During this time, I met some of my friends from before I went to Vietnam, and two of them said to me that our friendship was now over; "We don't want to know about it, we don't want to hear about it, just don't ring us, don't contact us, that's the end, we don't want to know."

The other thing that I did was to go to join what we call the RSL, the Returned and Services League, which is like your VFW. They told me that they didn't recognize Vietnam as a war zone so I couldn't join. They just said, "No, not interested. You haven't been to war, we're not interested."

Home

I started looking for a job and there was no way that I could put on the application that I had been in the Army for two years, because I wouldn't have got the job. So, when I was asked what I had been doing those two years I just said, "Oh I've been around Australia doing odd jobs here and there, and seeing all the places I want to see." So, you just wouldn't put your Army service on your resume.

I had worked in the insurance business before I went into the Army, and I found a job in that field. But I was only there for a short time, because I had contracted hepatitis in Vietnam. I was in bed recovering for three months, and I decided that insurance wasn't for me and I'd like to carry on my service, so I joined what we call here the Commonwealth Police, which is now called the Federal Police. It was a job that I understood, and they wanted people who had been in the Army and had the discipline and had the training.

I was having problems with my marriage at this time. My wife really didn't want to know what I did over there. No one ever asked me about Vietnam, what it was like, or what I did. I didn't know what was happening to me at the time, but, finally, my marriage was dissolved after 13 years. We had three children, three boys, and I am positive, without a doubt, that Vietnam and PTSD were the causes of my divorce. I would have outbursts of anger at small things, switching moods from being happy to saying nothing and just sitting outside by myself. I used to snap really quick. I really didn't understand what was happening to me.

Three years later, in 1983, I married my wife Robyn, and we have been happily married ever since. She knew I was a Vietnam veteran but I was reluctant to tell her family. But her father, Bruce, was a World War II vet, and when I met him I said, "Listen, I was called up for National Service and was in Vietnam." He said, "Good, good to hear," and then I felt very welcome.

There was a Vietnam Welcome Home parade held here in 1987. I never would have gone to that Welcome Home Parade unless my father-in-law absolutely insisted that I go. He told me it would do me a world of good. "You'll catch up with old friends, mates from the Army that you haven't seen for years." So, it was on his insistence that I went. He said, "This is recognition for you guys, the people have now forgotten, they want to see who you are, they want to give you a welcome home so you need . . . you absolutely need to go."

I had a real, real hard time with it, because I was still afraid that there would be people there that were going to do the same things that they did when we came home initially. I was afraid that there would be groups or something along the march route that that may cause a problem for us. But everything went well with the parade, and we've had a strong movement here ever since with our Vietnam Veterans Association.

I kept in touch with a friend from high school named Phill Evenden. We played cricket together. He was also called up and went to Vietnam as well. We are still in touch to this day. I was having problems and Phill knew the symptoms, because he had the same ones. He said to me, "You need to go and see this psychiatrist, and you need to sit down with them and let them know what's been happening to you." I was very reluctant to go, initially. I thought, "There's nothing wrong with me, no." Very reluctant to do it. Thanks to Phill, I finally did it, because I needed to accept that, yes, there was something wrong.

When I finally went, the psychiatrist explained to me that, in his opinion, when the Vietnam vets came home they did one of three things. They either became a workaholic, or became an alcoholic, or they committed suicide. He then explained to me why it was so late in life it all started to come out; it's just that you try to cover your tracks and then eventually you hit a wall, and I had hit that wall. I was diagnosed with PTSD. I still see him every four months.

Probably the last five years of my working life I had seven jobs, and that was part of my hitting the wall process. On my last job, I was going to punch someone out and that's when I went back to my psychiatrist, and he said, "Look, you've got to stop working, so I'm giving you four weeks off. Enjoy it, and then you go back to work." Well I had the four weeks off, and I wasn't back two weeks and I went back to my psychiatrist, and that's when we finally got things together. Within three weeks I got cancer in the groin. Then I had prostate cancer and had my prostate out. I am not sure it is due to Agent Orange, but I am sure I came into contact with it sometime. And there were other defoliants and chemicals sprayed on us besides Agent Orange.

When I was told grandchildren were coming along, I was absolutely terrified that there might be something wrong because of my exposure to Agent Orange in Vietnam. I've got five grandchildren, but fortunately only one has any issues; my last grandson, who's only one, was born with a club foot. I was absolutely, emotionally terrified that when they were born there might be something wrong with them, because it's the second generation that is affected from Agent Orange rather than the first, as I understand it. I was just absolutely terrified that there was going to be something wrong and that would have been my fault, really, at the end of the day.

I believe that wartime changed me as a person. I was a different person when I came home. One thing we have over here is a DVD called "Flight and Fight." It explains how the Army trains you and puts you into a fight mode, and then when you are discharged they don't take you out of fight mode. They left us in fight mode rather than go back to flight mode where a normal person was. Most people who come upon a confrontation would just walk away and not get involved. We stay in this fight mode because we've never been demobilized out of this area. That was one of the hard things to accept. And it does change you: you're different, you're not the same. There was no decompression time, you know. You're in Vietnam and ten hours later you're home. And then the welcome home was just, um, excuse my French, "Bad shit," really. To be honest, I would rather have turned the plane around and gone back.

Up until Vietnam I had been a churchgoer, but when I saw what was happening in Vietnam, I said, "There can't be a God because of the carnage that's happening here." It just begs the question in your own mind, "Why would someone let this happen to people?" To be honest, the only time I go to church these days is for a christening or a wedding. Other than that, I don't attend church, and haven't for many years.

I vividly remember when Saigon fell, and I just said to myself, "Why? Why? Why did so many die for something that is now meaningless?" It left an e mpty feeling in my gut at the time. It was gut-wrenching at the time; really, really gut-wrenching. I actually went more into a shell when Saigon fell. Even back then, when we loaded the wounded on the planes to come back to Australia, I would look back into the plane and see them being cared for and the coffins, and I just said to myself, "What a waste." It really was. And when Saigon fell I kept seeing the men with missing limbs and the coffins, and thought to myself, "What was the purpose?"

From my personal point of view, I thought my time in Vietnam was life-changing. It was a great experience. It certainly bonds people: the Army or services bond people together for a lifetime, so I think that's the positive for me on Vietnam. It has really helped me to be a different person in my life, having looked back on it.

And it's certainly an experience that we are able to pass on to our younger ones at school. In Australia, and particularly where I live, we have a roving bus that's called the Nambus, which is a bus fitted out with all the information on Vietnam. It has equipment and everything we had to sign up initially for the National Service. It also has a large book that has photos of all of those who were killed in Vietnam. We go around to schools when the bus comes up here and we speak to the students and tell them what life was like in Vietnam, and I think that makes a difference to those kids as well. They know about what happened, because it's part of their studies in school here. We do that throughout New South Wales when we can. It's good for us, as well. It's good for the guys to get up, but it's hard to talk about it in front of large groups. But we do it.

The only thing I need to add is that through Vietnam I was able to meet people, particularly Americans, and talk to them about their country and how their country has evolved, and where it is today. We were able to explain to them about Australia and how we live, which was not dissimilar in any way to the Americans. What we do here in Australia and what you guys do over there are virtually one and the same. That was a great experience, too, to be able to meet someone from an overseas country.

I met some young American Afghanistan veterans in Chicago recently and went to say, "Thank you" to them for their service. I mentioned that I was a Vietnam vet and was pleasantly surprised when they also said, "No, thank you for your service." Once again, I felt accepted.

Corporal Mick Cassidy

Royal Australian Army Ordnance Corps

I was born in 1949 and have lived my entire life in Brisbane, Australia. I was married in 1968 when I was 18, and divorced in 1982. I have two sons from that marriage. I have been in a relationship with my de facto wife for 40 years. She has a son and daughter from her previous marriage.

I joined the Royal Australian Army in 1969. I was 20 years old. I decided to join because I needed a change of life, and I thought I would end up getting drafted, anyway. I was assigned to the RAAOC, the Royal Australian Army Ordnance Corps. My military specialty was ammunitions and weapons.

I volunteered to go to Vietnam and was sent there in 1970. I spent my year at Vung Tau and Nui Dat. I returned to Australia in 1971, but remained in the Army until 1981. I was promoted to corporal in Vietnam.

Coming Home

I came home with a few soldiers from my unit, along with others from other varying units, so unlike the infantry battalions, who would come home en masse. I left Vietnam out of Vung Tau. We flew on a C-130 Hercules transport. There were lots of diggers from different units on board. When we were due to take off, the base at Vung Tau was being shelled by mortars, so we had to stay put, on the plane, until the shelling stopped. Lots of silent prayers were said on that morning, and we got in the air about an hour later.

The trip to Australia was very rough, very long, and very noisy, it took us 22 hours in the cargo hold of a Hercules transport plane. We made one stop at Darwin, in far north Australia, for some tucker, and then back on the Herc. for the remaining trip to Sydney. I had a mixture of emotions at this point: happy, excited, numb, and tired.

When we arrived home, we had blokes from many states of Australia, so we dispersed into different groups, were processed, and given our tickets to fly back to our home state. I think we were allowed to change into civvies, so we didn't stand out too much, but the fact that we were all carrying olive drab duffel bags and had very short hair made it fairly obvious we were soldiers. The civilian clothes weren't real good camouflage, and we did cop a bit of abuse from the long-haired louts that were there to make their presence felt. The general public was okay; some others were not so accommodating. We were subjected to some abuse mainly from the right-wing rabble, and long-haired university students.

When we landed in our home town, we had to find our own way home. I don't think I had notified my wife before I left Sydney, so it was a surprise for her and my young son, who was three years old. I should have let her know I was coming, though, as she wasn't home when I got there.

Initially, it was strange being home. There were a lot of things that I struggled with, mainly change of lifestyle. I was comfortable for a while, but I think anxiety was always hovering around. My family was fairly supportive, and most of my friends were military, so we all shared ups and downs, and we were very supportive of each other. But some of my friends were a bit apprehensive as they were anti-war. Others were okay, and a few just didn't want to know you. But there was a large military community where I lived, so I just socialized mainly in that group.

However, I soon started to drink heavily, and would spend a lot of time and money at the boozer with mates, and ignoring my family. I had bouts of anger and my marriage eventually failed. It failed due to my drinking, not having enough money to pay bills and buy food, so my wife simply had enough, and our marriage ended bitterly.

Home

I remained in the military for another ten years after I came home from Vietnam. Being around others who were over there helped.

I had a certain degree of problems after the war, mainly anxiety. After I left the military and went to work in the private sector, I started having problems with authority. I could not adjust to the civilian way of doing things. I became very frustrated and insubordinate, and often clashed with other staff and, in the end, I was sacked. This was a bit of a blessing as I ended up getting a job back with the military, but in a civilian capacity.

After a few years, we had a Welcome Home parade, similar to what you blokes had in the States, I met up with some mates who I served with, and over a few beers after the march we all got talking. I was surprised to hear them speak about similar sorts of things that I had experienced since coming home. I eventually went to see a shrink at The Keith Payne Unit, which is a psychiatric facility here in Brisbane. He diagnosed me with PTSD and survivor guilt. I am on a TPI (Totally and Permanently Disabled) pension, and am still seeing the shrink, been seeing him for many years.

It all seems so long ago, but there is probably not a day that goes by that I don't think about it in one way or another. I still have flashbacks, and I often have panic attacks.

Chapter 13 - Reconnecting

One of the hardest things for Vietnam veterans to do is attend their first reunion. Almost universally they feel nervous and very apprehensive. Will they know anyone? Will anyone remember them? Will it bring back memories best left buried?

My first reunion was in 2010. There were 16 of us. I was very reluctant to attend, even though I was the one who organized it. I was about to see some old friends that I hadn't seen for 40 years. Would it be awkward? That was a long time ago. Would we have anything to talk about? When I arrived, I walked into the bar and there they were! We laughed, hugged, and felt the years melt away. Time had been turned back. We were young again. Nothing had changed. It was like yesterday when we were together back in Vietnam.

After things calmed down a bit, my old buddy, Bill Fesus, came up to me and said, "I have something for you. Hold out your hand." I extended my hand and in it he placed a black rosary. I stared at it, puzzled. What was this about? He said, "You don't remember this, do you?" I honestly told him, "No, I didn't remember it." He told me a story I had forgotten a long time ago. He said, "There was a point in Vietnam where you were mad at God. Your grandmother gave you this rosary before you shipped over, and you were going to throw it away. You knew I was religious so you asked me if I wanted it. I have been saving this for you for 40 years." I was stunned. My heart felt like it skipped a beat. That rosary is now one of my most prized possessions. We made friends in Vietnam who became brothers. Reconnecting is something that you can't put off, no matter how hard it seems. The clock is ticking and our time is getting short.

For a couple years after I came home from Vietnam I stayed in touch with two or three buddies with whom I had served. My buddy Joe Kaminski was one. He was from Brooklyn, and in 1975 I was in New York on business and we were able to get together for a couple days. We had a great time. One of his buddies was a cab driver, and he drove us all over Manhattan and showed me the sights. But, after that, time went on and life got in the way. We talked on the phone every few years and we lost touch with each other. In 2010, after our first reunion, I decided to find Joe again.

My daughter was living in Manhattan, and in 2011 my wife and I drove to New York to spend Thanksgiving with her. It was time to scour Brooklyn and find Joe. I found three Joe Kaminskis living in Brooklyn, but none of them had a listed phone number. The day after Thanksgiving, armed with my GPS, I decided to visit all three and hopefully find him. The first two stops were a dead end. I had one more chance to find him. I walked up to the door of a three-flat and rang the bell marked Kaminski. No answer. I rang again, without a response. I went to my car and came back with my flashlight and proceeded to hit the door real hard three times. The buzzer sounded, letting me in. I walked in the foyer, and a man on the second level stared down at me and asked how he could help me. I told him I was looking for a Joe Kaminski that served in Vietnam. He stared at me for a few minutes and finally said, "You don't remember me, do you? I am Joe's friend, the cab driver that showed you the city in 1975!" Oh, my God, I couldn't believe it! He invited me up and his wife made tea. Joe was at work, and he called him on the phone and told him I was there. Joe rushed home and we had a great visit, just like 1975. Joe swore he would come to the next reunion. But, three months later he was struck with cancer. Agent Orange reared its ugly head, and Joe died a month before the 2012 reunion. At least I was able to see him one more time. Another casualty of the Vietnam War.

In 2013, I found my old first sergeant, Marvin Bentz. He was a great guy and always looked out for the troops. He didn't really remember me but was really excited to be contacted. I put him in touch with our old CO, Joe Matos. Marvin was really excited about attending the next reunion in Washington, DC, from October 16 to 20, 2013. He couldn't wait to see Joe Matos and all the guys again. On September 3 I received a call from his daughter. He had suffered a stroke on September 1 and passed away.

Everyone finds it difficult to attend their first reunion. Nervousness and apprehension can get the best of you. As you walk to the door of the event that old familiar tightening of the gut will grab you. Keep walking, because when you see the buddies you served with 50 years ago, all those feelings will disappear. You will be young again, and you will find that no one has changed. Do it before it is too late: you will never regret it.

Marine Vietnam Vet: I was very apprehensive about attending my first reunion. My first reunion was in Montana. I flew there from Ohio, too far to drive. Hell, I haven't flown since 1993—OMG, how things have changed, even the first-class seats were smaller than I remember! The last time I flew first class was when I came home from 'Nam in 1967. There were only three Marines I recognized from my time there, but the older salts made me feel right at home. It was an experience I will remember the rest of my life.

Navy Vietnam Vet: I've gone to two reunions so far and am looking forward to the next one, when I will meet up with the guys that I served with in 'Nam. I hooked up with two guys that I haven't seen in 51 years! It was great meeting them again.

Army Vietnam Vet: I still don't talk about being a Vietnam vet. I go to unit reunions, but I feel it's nobody's business. They weren't interested back then and I don't like to hear that, "Thank you for your service" now—it comes off as a real shallow statement.

Army Vietnam Vet: What a great time we had at our 45th reunion. It was truly an honor to serve with them in Vietnam.

Army Vietnam Vet: I attended a reunion in Washington, DC. We stood together as brothers and had an incredible ceremony at the Wall. Seeing the Wall for the first time was very emotional for me, and I was very glad to share those moments with my fellow brothers. I'm glad I attended that reunion, seeing more guys that I have not seen for over 40 years, and meeting new guys that were there before me.

Army Vietnam Vet: I remember the Chicago Homecoming Parade. I think that was 25 years ago, I don't remember the year. I remember driving to work, hearing about it every day on the radio; "There's going to be a homecoming for Vietnam vets." And then, suddenly, I realized, "You know what? I'm going to take a day off and go." And I did. I got out my uniform (what would fit on me) and my boots, and I went down there. And it was just like you walked into a football stadium with people you knew. And I didn't know anybody that was there; I was just saying hello to everybody. We marched out in the units we had served in. People were cheering, like, and I felt, "Wow, I've never walked down a street where people cheer you." The steel workers at one building were throwing six-packs of beer, so we would have something to drink while we were marching. It was very nice. Very nice.

Air Force Vietnam Vet: Air Force weathermen in I Corps were few and far between, so no reunions of our little operating location. But last year, out of the blue, I got an invitation from C/Co. 158th Aviation Bn. the Phoenix, to their reunion. Members of the Phoenix were my best friends at Camp Evans, and although I stayed in touch with only one or two of the guys, and only sporadically, over the last 44 years, somehow they found me and I was the only non-Army guy there. What an honor. My first and only reunion.

Chapter 14 - Donut Dollies

The American Red Cross had a presence in South Vietnam from 1962 until 1973. As the war increased in intensity and troop concentrations grew, more and more Red Cross workers arrived to provide assistance to US Military forces and many refugee hospitals and camps.

The Red Cross provided three types of service. The first was working in US and civilian hospitals, providing medical services where needed. Another was to help military servicemen and women with their personal and family problems. This included offering assistance for troops who had to get home on emergency leave due to the death or serious illness of a family member. The Red Cross handled 2,168,000 emergency communications between military personnel and their families during the war.

The third service was the Supplemental Recreation Activities Overseas (SRAO) Program. The Red Cross recruited female college graduates to work in this program. They were called Donut Dollies. These dedicated young women could normally choose to go to Korea or Vietnam. They received two weeks training in Washington, DC, before going overseas. They made the long flight to the Far East with American troops on chartered airlines. Once there, they received their assignments and began their work, entertaining the troops and bringing a touch of home.

The Donut Dollies manned recreation centers at various bases throughout Vietnam. These centers would have books, coffee, Kool-Aid, and music and other activities to entertain the troops. From these bases, they traveled throughout their assigned area of operations, visiting troops at outlying helicopter bases, fire support bases, and in the field. They traveled by Jeeps, trucks, helicopters, and planes. Many of these trips were very hazardous.

They were always smiling, playing games, handing out "short timer" calendars, passing out cigarettes, candy, stationery, and other items the troops always needed. You could always recognize them in their powder blue uniforms.

A total of 627 volunteers served in the SRAO Program. It is estimated that over the course of the war they logged over 2,125,000 miles traveling around the country entertaining the troops. They lived on base with American troops and were subjected to rocket, mortar, and sapper attacks. Their convoys received sniper fire and the helicopters they flew on were occasionally shot down. Five Donut Dollies were killed and more were wounded and injured.

The important thing to remember is that they didn't have to be there. They all volunteered. They received no medals, rewards, or benefits for their service. They were exposed to enemy fire and Agent Orange. Today, many Donut Dollies suffer from PTSD and illnesses due to their exposure to Agent Orange. They get absolutely no help from the government, Red Cross, or VA. PTSD counseling must be at their own expense. Health issues due to Agent Orange and other herbicides must be treated privately.

They raised the morale of the troops everywhere they went. They came home with the troops they entertained and were faced with the same anti-war sentiment and adjustment issues that the military personnel were. They are Vietnam veterans without DD214s, and we love them all.

It was an awful, awful experience for me to see 50,000 people cheering at a football game when I had just left a war zone with Americans dying.

Melinda Koester Pendino

American Red Cross Donut Dollie

I graduated with an AB undergraduate degree in May 1969, from a somewhat elite, small "girls school," otherwise now known as a woman's college—Sweet Briar College, outside of Lynchburg, Virginia. I had spent three years at Sweet Briar with my junior year in Madrid on the NYU program. During my senior year, I spent time visiting my childhood friend, Jackie Pratt, and my high school boyfriend, Tommy Pendino, in Washington, DC. Tommy was stationed at a Navy base nearby. I saw protesters holding signs in DC. At school, I heard the daily body count on the "Today" show as I had a small television in my room. It all seemed so fruitless.

As I was the daughter of a Red Cross woman who served in both theaters of war (New Guinea and Berlin), and the daughter of an Army air corps lieutenant who sang "Up in the Air . . .", also serving in both theaters, in England and in Tinian, it seemed very natural for me to explore the American Red Cross while I was in my senior year at Sweet Briar.

I called the Red Cross and, almost immediately, two women agreed to an interview at the Presidential Inn in Lynchburg. That was the nicest motel in the area, with the rooms encircling a large swimming pool. My roommate had a car and drove me the 15 miles to Lynchburg. Her high school/college boyfriend and later husband was also from Columbia, South Carolina, and a West Point graduate then stationed in Vietnam. I had assumed Joe Jackson and Peggy Gibbes would be very supportive of the interview. They knew the Westmoreland family from South Carolina. Instead, Peggy talked about Joe's adamant objections to this interview. When we got to the Presidential Inn, I remember Peggy driving me around and around the pool, not slowing down to let me out!

But, finally, I got out. The interview was very cordial and a "done deal." I was now a second-generation Red Cross worker.

My parents spoke very little about any specifics of their war to my sister and me, except in light, anecdotal terms. My mother used to talk of the soldier who sang of "The Northern Lights and the Southern skies" in New Guinea, and my father recited he wouldn't eat "lamb, mutton, or goat," as that is the only meat they had in England. Daddy's job was in the officer's clubs and at the body bag sections; very different work sites, but both behind the front in England, but not so in Tinian. My mother worked as a social worker in the medical wards.

The war had really just ended in my childhood, and I marvel now at how little it was discussed with us. My mother had a number of confiscated German items she brought home, like a gun, plus some German porcelain. I have an oil portrait of my mother in my dining room that a soldier painted in exchange for cigarettes. It is quite good, I think. My mother had close Red Cross friends from her time served that I knew during my childhood and when I grew up. They were attractive, well-educated women. Daddy did not seem to connect with his service coworkers. He put it behind him, talking very little about it until his death at age 75 in 1985. It probably would have been healthier to talk about the dark side of what he did, but he was an Ivy League-educated Wharton Business School graduate who did not ever talk about feelings.

In September 1969, after August training in hot Washington, DC, I was stationed at Bien Hoa with the 20th Engineer Corps, then at Da Nang with the Marines, and lastly at An Khe with the Funky Fourth.

My time in Bien Hoa was longer than most duty stations (almost six months), as we experienced the first death of a Donut Dollie while I was there. Hannah Crews fell out of the front seat of a Jeep when it hit a pothole in the dirt road, and she hit her head. The pin to the front seat hadn't been secured by whomever got in the back seat. She was injured the night before I arrived in Bien Hoa, and I met her in Long Binh at the hospital. The first time I saw her, she was beginning to fail, and soon after that she went into a coma. She died maybe a week later. There was great emotion and controversy about her death. Whose fault was it? Why weren't her parents allowed to fly into Long Binh, as her mother was a statewide Red Cross volunteer? How/where/when do they care for her in ICU in Long Binh, or then transport her, as women didn't frequently die in Vietnam? We tried to carry on our routes and duties, but it was nearly impossible. The powers that be thought the unit would "heal" better staying together, so we did, and only replacements for those going home came into Bien Hoa.

My second duty station was in Da Nang, at Freedom Hill (I think it is called), living initially in a gorgeous villa downtown, riding in a deuce and a half daily to the Red Cross center. Then, for "security reasons" we

were moved to a trailer right on the beach! The beach at Da Nang is gorgeous, and we had very nice quarters. While at Da Nang, we had a two-night overnight in Quang Tri. That was the toughest challenge of my year in Vietnam. It was where we went to the makeshift hospital, and the nurses asked us to hold the hands of patients being medevac'ed to Japan. We were told that usually the guys didn't know the extent of their injuries, and we were there to be positive, caring friends. It was tough. Body parts were missing, and the outlook for many was grim. We also went to the body bag tent regularly, to talk to the guys as they marked and identified the bodies. We used an outhouse in Quang Tri, but that seemed easy compared to the baggage of war we saw and experienced there. The trip to Quang Tri was by helicopter, of course, over the gorgeous blue sea, up the coast from Da Nang. For some reason, I seemed to go frequently as it was important to me. It was at a program outside of Quang Tri that a Red Cross worker found me and told me of the passing of my dear grandmother. He offered me the opportunity to go to Omaha, Nebraska, for the funeral, as my cousin stationed in

Germany was going. But, realistically, it would have been two days just to get out of Vietnam and lots of favors to ask due to that remote location, so I declined.

My last duty station was at An Khe. While there, the second Donut Dolly was killed in Cu Chi, but the circumstances were not readily known then. The unit director was away on leave, so I had to beg for the two girls in my unit who had just arrived in-country with the girl who was killed to attend the service at Cu Chi. I didn't go. Susan McLean Bradshaw knew the girl only a few weeks in DC training, but her grief was overwhelming. I am glad the general was kind to us for that transport. An Khe was gassed and had incoming a good bit. Hot water was sporadic for showers, and we occasionally had to eat C-Rations, so I used the cookbook from McIlhenny Tabasco. I continue to buy that brand. Interestingly, my husband went to Louisiana to the plant, but no one would admit to that program. Strange, as it made the food palatable to me, and I became a fan.

Coming Home

It was a solitary trip home to Tampa, Florida, from Saigon, where we all had to go to process out. I might have had a purse packed, but I remember that I wanted to carry my ditty bag purse as a proud reminder of the war. I remember carrying my ditty bag through the DFW Airport. My uniform was very worn, as it had been washed on rocks by the mama-sans for many months. The trip home seemed quick compared to going over. When I got home I went to bed. I slept for days, but I was freezing all the time. This body temperature change was very unexpected, as Florida is warm like Vietnam, and it was October, but I wasn't used to air conditioning, and my home, church, stores, were all so cold I shivered for weeks. I went with my parents, and a guy who didn't serve in Vietnam (Navy Reserve) to a football game at the new stadium in Tampa called the Big Bolero—Florida vs. Auburn. It was an awful, awful experience for me to see 50,000 people cheering at a football game when I had just left a war zone with Americans dying. I wanted to scream out about the guys dying so they could have the privilege to see this game. But I was "raised right," and my father had thought and planned for this game as he knew I loved football. I didn't want to hurt his feelings, but that game was pure torture for me, and I don't think I ever dated that guy again. No one understood.

At Christmas a few months later, I got a job at the local department store in the china department. That was all I wanted to do. Then, a few months later, I got a job as the activities director at a local adolescent psych. unit at a Catholic hospital in Tampa. After that, I was hired as a child abuse social worker/investigator, until my daughter was born in November 1976.

Home

As my first husband didn't serve, I spoke of Vietnam very little. A neighbor (and church member) was a Donut Dollie. She brought me a plant when I moved in, not knowing me. My aunt was visiting from Omaha, and I was at work. Somehow my aunt found out Mary Jo had been a Donut Dollie the year before I was there. We mentioned it to each other occasionally but not much. Another Donut Dollie, Linda Bryant, organized our first get-together of DDs in Orlando. Linda worked for the Veterans Administration, but, more importantly, had some of the databases of the Head of the DD program in Washington, DC, so she has truly been a tremendous asset to our program and subsequent gatherings.

I married my high school boyfriend in 2001, after my first husband had died. Tommy wrote a poem about me visiting him in Vietnam, which I did get to do, thanks to a great Navy admiral. That is an amazing story, as Tommy was an enlisted sailor on a destroyer in Da Nang Bay. But I broke no rules!

Before I married Tommy, I didn't really talk much, if ever, about Vietnam, and few coworkers or friends knew of that year. Tommy revered me and my service. He gave me a gorgeous diamond, ruby, and sapphire American flag one Christmas, "For your service!"

There is a newspaper article about a DD reunion we had in St. Petersburg that includes my picture at a fire base. I, like my parents, didn't talk about that year of service much. But it doesn't nor did it define me or my life, other than I experienced far more than I could ever repay. The guys didn't want to be there, so I thought it was only right that I, like my mother, answer the call to serve. Hopefully, I have continued to give back to my community. As an aside, my daughter called the American Red Cross in DC after 9/11 when she was in second year law school. She was ready to sign up! It appears there is no similar Red Cross DD service now, so she continued in school.

I am glad I served a year in Vietnam during the time of high troop levels. Protesting the war in the States seemed to serve the drafted grunts very little. Those guys had no choice. Many friends and acquaintances from Tampa died in Vietnam, including the brother of my husband's best friend, Joe Turnbull. He was like a little brother to my husband.

I had a hard time going to the Wall, and only went in the late 90s when a Florida Lottery coworker, Harvey Malter, literally dragged me there when we were in DC at a lottery conference. I had been to DC after the war, but I just didn't want to go. I am so glad I saw it. I looked up many names, including the name of my bridge partner, Lt. Robert (I called him Bobby) Wehunt, who the next morning died piloting his chopper outside of Bien Hoa. He was an only child of an older North Carolina couple. I only wish I had reached out to them when I came home, but it is a regret.

This is my story. I hope it is helpful. It is not a dramatic portrayal of my year but, rather, a chronological review of my year that is a little different than anyone else, as each story is. God bless you for this 50th Anniversary. To each of us in memory of those that didn't come home.

Penni Evans
American Red Cross Donut Dollie

I was with the American Red Cross as a Donut Dollie in Vietnam from March 1970 to March 1971.

I attended Colorado Women's College in Denver from 1966 to 1968. I then transferred to Sonoma State, which had lots of strikes, armband protests, etc. I graduated college in three-and-a-half years in order to go to Vietnam. I joined the American Red Cross because of JFK's speech: "Ask not what your country can do for you—ask what you can do for your country." I thought about the Peace Corps, but they wanted foreign languages which I was not good with. I did not believe in the reasons we were in Vietnam, but I did believe in the men and women who served there. So, I joined.

After my training period. I flew from Washington, DC, to San Francisco. I guess that emotionally I was mostly excited. I then flew from Travis AFB to Hawaii, Guam, then Saigon.

I was a recreation worker. We provided recreation centers at larger bases and took out silly game show-type programs to the units at fire bases and in the field. We went anywhere where we were requested by the military to provide a presence. It was a touch of home.

I was at Cam Ranh Bay Air Base, where we worked at the Naval Air Facility Cam Ranh and with the Seabees. We traveled all the way out to the border, Ban Me Thuot and Ban Don being scheduled overnight trips. I then was sent to II Field Force in Long Binh, where we worked with the 11 ACR, 199th LIB, and some of the 9th Infantry Division out of Dian. After Long Binh I went to Cu Chi to reopen the unit which had been closed after a fellow Donut Dollie, Ginny Kirsch, was murdered in 1970. I was originally scheduled to go there the day she was murdered, and I had my choice of an afternoon flight or a flight the next morning. I chose the morning flight at the urging of my Caribou friends, so I could go to a goodbye party the Caribou crews gave me the night before. Our unit director was called about 5:30 a.m. to tell me not to get on the plane, to stay put. No other information was provided, so we did not know if something had happened during an attack, or what.

We did open the unit there but closed it as the 25th Division was going back to Hawaii (but most of the men were sent to other units to finish out their tours). We opened the unit at Quang Tri, where the 1/5 Mechanized Infantry was stationed. I was there for the buildup and beginning of Lam Son 719.

I had a hard time dealing with the hospital visits, but it was so important for the men, so I had to hide my feelings, not of horror but of wanting so badly to make it instantly better for them. Handing out short-timer calendars to men going out on patrol was hard, as I wanted just to grab them and make them stay.

Coming Home

When my tour was over, I left Quang Tri on a medevac plane to Cam Rahn Bay. While waiting to load I held IV poles for the wounded.

Another Donut Dollie and I took a commercial plane to Okinawa. As we lifted off from Tan Son Nhut in Saigon, we yelled! We were going to go scuba diving, but the weather was horrid. So, we decided to go home. We had a 27-hour standby and finally caught an Air Force C-141 to Hawaii, then onto Norton AFB in California. I took a bus to San Diego, where my sister and her husband picked me up at about 4:00 a.m. One of my sister's neighbors told me I could not have been in Vietnam because I did not have a tan! She obviously had never heard of the monsoons!

I visited a friend in Los Angeles who took me out for an Arby's sandwich. Walking about looking at the earthquake damage startled me, as I had only heard reports of it while at Quang Tri. My friend was great. He was a Tonkin Gulf Yacht Club veteran on the USS Bainbridge.

I flew to my family home in Indianapolis. My daddy was transferred from Whittier, California, to Indianapolis right after I graduated from high school. I landed in a snow storm, and my mother was there to greet me. I think she was disappointed I was not in uniform, but I did not want to wear it traveling. She took me to Baskin Robbins for a mocha almond fudge ice cream. My parents' friends were very supportive of me, and watched my slide show (first time I had seen the pictures I had taken since they were sent home).

The first night home I woke myself up screaming! I have no memory of the dream. My mother wanted to comfort me but my father told her to let me be. I just needed time. We only talked about it once in the morning.

Home

Daddy had to give a speech at a college outside of Chicago and invited me to come. He was talking about the Naval Reserves. He was a career naval officer and wore his uniform, and introduced me as his daughter just home from Vietnam. We were booed! He was taken aback and hurt. He was so proud of me.

I backpacked in Europe, leaving in early April, and was there six months. That eased my transition a lot. When I got back, I applied for the Navy OCS but was rejected because their quota for women officers was filled. I reapplied later in the year, but was turned down again.

I did not go back to my hometown in California. Indianapolis was not home either. I missed being in Vietnam, though I could not wait to leave. I learned not to tell people I had been in Vietnam, as the reactions were not positive. I was told I betrayed my generation and that I supported baby killers, etc. As a friend said, the easiest way to clear a room was to say you had been in Vietnam.

I felt disconnected from life. I sought out counseling and was told Vietnam was not an issue. This was in the mid- to late 70s. I went back to school and got a master's in human behavior, hoping to go into training. But, I ended up in human resources. While working at National Cash Register (NCR) in 1977, I was taking papers to payroll and they were putting up movable walls using a gas-charged bolt gun. They set off two and I hit the deck. The lady in payroll is screaming, "They shot her, they shot her!" That was the first time they knew I was in Vietnam.

One of the recruiters at NCR was an Air Force vet, a casualty notification officer. The head of security was a three-tour Marine, and he and I talked a bit, but I could not keep up with his drinking. In 1982, at a Fourth of July party in Whittier, we were walking back from the Uptown area when someone let off a bottle rocket. I hit the dirt and a lady who did not know me remarked that she had never seen anything like that outside a movie.

I was still feeling disconnected, a step out of time, and could not figure out why. Vietnam never really entered my mind then.

In fall 1983, I moved to Santa Rosa and went to my first Donut Dollie reunion in San Francisco. It was there that so much came together, seeing friends from Vietnam, talking and finding I was not alone in my feelings. Some wonderful ladies walked point for us and helped us along our journey home. I found a grass roots veterans center that did not care that I did not have a DD214: they accepted me as a client, and, most importantly, as a Vietnam veteran. I went to the Wall for the Memorial Day weekend in 1984 after a Vietnam Women's Veterans reunion in New York City. Then I left to visit my parents, who were living in the British Virgin Islands. I tried talking to my parents about the reunions and my feelings (which were pretty raw at that time), but my mother could not understand. She lived through four years of war, what was one year? And Daddy said just to let it go and get on with life. I wrote them a letter when I got home explaining why I could not tell them about Vietnam. It arrived the day Mother was killed in a car accident on the island.

I began to speak at colleges and high schools, using the journals I kept as part of counseling. I also got involved with the local Vietnam Veterans of America chapter and with the Vietnam Women's Memorial Project. I was on TV, radio, interviewed by papers and magazines. I was interviewed for a book, "A Piece of My Heart: the Stories of 26 American Women who Served in Vietnam." I was part of a Sonoma County group that was flown to Madison, Wisconsin, for a PBS show, "A Program for Vietnam Veterans, And Everyone Else Who Should Care." I went to Chicago for the Welcome Home parade, then Houston's Welcome Home and Stand Down. I participated in local parades as well. All these events made me a part of the Vietnam veteran community. Out of my journals, I began writing poetry about Vietnam. One of my poems is in "Visions of War, Dreams of Peace." The sisterhood from Donut Dollies, Army Special Services, USO, and nurses and enlisted WACs, has been so positive. And the welcome by the brothers has been awesome.

I was housesitting in La Jolla when Saigon fell. I watched the TV in shock, in tears (which, I did not cry easily), and I called my parents. The two high school girls I was "baby sitting" were so concerned. One found me from an article in the paper in the mid-1980s and told me how impressed she was from that night. And she said she was proud to know me.

I believe my PTS (I hate the "disorder" part, as to me it is a normal reaction to a very abnormal situation) began in the early 1970s. And I found The Flower of the Dragon (a grass roots vet center in Cotati, California) in early 1984. That was a great help.

I have small nerve neuropathy and gastroparesis. I had internal bleeding in 1974 which led to hospitalization and ICU, but they never found the site or cause. Had digestive issues as well, and had my gallbladder taken out in 1988. It just went into spasms for no known reason. In the late 1970s I had sores on my scalp, like pimples, but which were crusted with yellow scabs. I still get flare-ups. My muscle pains have been with me since the mid-'70s. Are any of these from Agent Orange? I believe so. But we civilian women asked to be included in the Agent

Orange settlement and were denied because we were civilians. Too many of my women friends who are Vietnam veterans have health issues: Parkinson's, breast and uterine cancers, muscle weakness, and neuropathy. And two nurses I know have died from what I believe are Agent Orange medical issues.

My experiences in Vietnam were life-changing. As one man said; it was the best of times and the worst of times. My experiences gave me skills to think on my feet, think and act outside the box, not to sweat the small stuff, appreciate the little things, and to honor and respect people from all walks of life and areas of the United States. The American fighting man is a very special person. It was an honor to work with them all.

My perspective of the world and mankind changed, and not always in a positive way. I value solitude more today than I did before Vietnam, and I value friendships made from that experience, both in-country and all the years after. I am more tolerant in some respects but have little respect for people's petty issues (which to them are the end of the world).

It has been the reunions, the speeches, the parades, and the community of veterans over the years that have caused the positive feelings to outweigh the negative. I still have startle reactions, I still dream of Vietnam, of being taken back against my will, of being trapped, but I have learned to look to my life when I have the most negative dreams and see what is out of control or bothering me. I do enjoy the good dreams of friendships and beautiful beaches, and the sense of peace. Several friends have returned to Vietnam, some more than once. I no longer feel I have to return. That year in Vietnam formed me into who I am today. It helped me in my work, especially thinking outside the box. The support I received from my coworkers as I worked my way home from Vietnam also was so positive. I lived in Albuquerque in 1990, trying to find work after being laid off in Santa Rosa, CA. I found a veterans center and was accepted, and then invited into a combat veteran group. We were together when Desert Shield-Desert Storm was building, and helped each other through our feelings which came up because of that. I have been fortunate. In Peace.

Barbara (Bobbi) McDaniel Stephens
American Red Cross Donut Dollie

My younger brother, John McDaniel, served in Vietnam as an Army helicopter pilot with 20th Engineer Brigade at Bien Hoa, from February 1969 to March 1970. He arrived in Vietnam a little over a month after I did (during a ground attack marking Tet on Bien Hoa AFB). Against Red Cross rules, he was assigned to the unit responsible for our quarters, mess, and safety, right across the street from our hooch in Bien Hoa. My unit director kept this info from our headquarters in Saigon for three months, until 20th Engineers published an article in their newsletter about John rescuing two Donut Dollie's from a downed chopper; one being his sister. Of course, our headquarters learned of it, but didn't reassign me for a month to give a photographer from stateside time to do some publicity shots. I was then reassigned to Da Nang. I think John only talked about his experiences there to me and his roommate from flight school, who he reconnected with after he returned home. He has had half of his lung removed due to Agent Orange and is considered 100 percent disabled (Agent Orange and PTSD). He is living as a semi-recluse in St. Augustine, Florida.

I first learned of the Donut Dollie program from a Red Cross volunteer who also was doing volunteer work at the Los Angeles Free Clinic. I had developed an interest in the Far East during college from an Ancient Chinese/Japanese History professor. At the time, I was a beach bum/flower child in Malibu, California, and had been accepted by USAID for work in Vietnam, but they were in a hiring freeze. I contacted the local Red Cross office to learn more. They were in a recruiting mode for all three programs in the Far East: Field Directors, Service to Military Hospitals, and SRAO (Supplemental Recreational Activities Overseas). So, I joined the SRAO program in January 1969. After the interview process and completion of extensive paperwork, I was told to report to Red Cross Headquarters in Washington, DC, January 18, 1969, for training and shots. My dad flew out to California to help me move my belongings back to Ohio. He spent most of the drive across the country trying to talk me out of going, but he had also raised me to be independent. I convinced him (and myself) that I wouldn't be in danger, and that the men being drafted and sent there deserved our support. If it was okay for brother John to be sent there (he was finishing flight school in Texas, and suspected he would be sent to Vietnam), it was okay for me to go. My dad, mom and three younger brothers (John was still in flight school) all came to Washington with me, and Dad's last words to me before boarding my flight to Travis AFB were, "You'd better follow every order given to you! Stay safe."

Our working title was Recreational Aide, Supplemental Recreational Activities Overseas. We had two types of activities: Programming to remote locations (LZs, FSBs, Evac Hospitals, rice paddies,), and actual Recreation Centers on some bases.

Programming consisted of developing approximately one hour of different activities related to one theme, that was portable in a large olive drab canvas bag by two DDs, usually transported by helicopter, Jeep, or deuce and a half. These were usually adaptations of game shows or board games. Some had more physical elements like whiffle ball or football. The purpose was to remind the troops of home and give them a short respite from the daily grind of war. We were assigned to a unit for the day and usually made eight to ten stops at their discretion, and could include serving a hot meal, delivering rations, or just chatting with the men. At times, a whole unit (four to 13 DDs in a unit) could spend a day at a stand down or refitting of a unit. The largest that I participated in was 600 Marines playing a Red Cross football game (DDs were the refs with whistles and striped hats.) Wish we had a video of that!

If our unit had a recreation center we developed at least one special activity per day; e.g., a pie-eating contest, pool tournament, fashion show, holiday themed parties, etc. Each center usually had musical instruments, foosball, a pool table, card tables and games, ping pong, paperback books, stationery, etc. and always Kool-Aid.

When I received my orders to Vietnam, I was elated and excited to see that part of the world. I knew that what we were doing there would have an impact on my world in the future, and I wanted to be able to understand those consequences. However, arriving at Tan Son Nhut in the dark, being herded into a school-type bus with metal bar windows and two ARVN soldiers with rifles for the drive to our Saigon quarters did give me pause. We were required to take two one-week R&R's while in-country, evenly spaced out (probably to decompress), so I had an opportunity to visit Hong Kong on the first one (June '69) and Australia (November '69). While in Australia I was hosted by my Danish grandfather's brother's family in Sydney and the Blue Mountains, which helped alleviate my trepidations of spending my first Thanksgiving and Christmas away from home and family.

The Red Cross liked to transfer you every four months to different locations. I was told by the Saigon Headquarters that this was to give us different experiences and prevent us from getting too attached to specific units and/or men. I served at Bien Hoa (February '69 - June '69). This was a mobile unit only, no recreational center. The units served included: 20th Brigade Engineers, 1st Cavalry, 101st Airborne, 82nd Airborne, 11th Armored Cav, Bien Hoa Air Force Base, and the 11th Aviation Bn.

I served in Da Nang (June '69 - October '69). This was a mobile unit and large recreation center at Freedom Hill. The units served included: 3rd and 1st Marine Divisions, 199th Light Infantry, 1st Signal Brigade, 3rd Marine Air Wing, US Coast Guard, US Navy, USN Sanctuary Hospital Ship, Da Nang Air Force Base, and 1st Cavalry Division.

My last assignment was at Cam Ranh Bay (October '69 - February '69). Here we had a mobile unit and recreation center. Units served included: 1st Logistics, 5th Special Forces, 1st Signal Brigade, and U.S. Navy Seals.

Coming Home

Donut Dollies had the same superstitions as the guys about "short timers," so the one leaving country would be given the first choice in the last week of runs and recreation center schedules. I was the program director at Cam Ranh Bay, Army, so it was important that I train my replacement (the last week that I was in-country) regarding any new runs and the logistics involved for the unit. Three days before I was to leave, we went by Caribou out beyond Ban Me Thout to a Special Forces/Montagnard outpost. Unfortunately, weather prevented the plane from returning to get us before nightfall. I tried to hide my rising short-timer anxiety regarding where we were, the obvious lack of accommodations for two women, and the safety of the men around us. My partner let it be known this was my last run, so the men got their short-timers superstitions going. After sharing some C-Rations and warm beer dropped off that morning with us, they settled us in the only bunker. I did not sleep a wink, and after sharing more C-Rations and coffee the next morning, our Caribou was able to land, and we made it back to Cam Ranh Bay safely.

Typical of military orders, my DEROS from Vietnam to Travis AFB was one week before I was due at Red Cross headquarters in Washington, DC. On the flight from Saigon to Travis, one of the men insisted on sitting next to me. I had on my Class A uniform with name tag, and he asked if I was Bobbi from Da Nang. Thinking he was from a unit I had visited while I was there, I told him I had been in Da Nang from July until the end of October. He proceeded to tell me that he had won $50 by betting on me. This was so unusual I asked how he did that. He said that he was a radio operator for a unit in Pleiku and had bet that I wouldn't leave country with my boyfriend from Phu Loi in July. He was referring to a sergeant, not my boyfriend, who had called me every day by field phone from the time I left Bien Hoa until he left country the end of July, trying to convince me to go home with him. The radio operator told me that he had broadcasted our conversations to his mess hall and that they had done a "pool" on whether I would give in and go home early or not. Needlessly to say, that blew my mind, and we both had a good laugh.

Due to the conversation about the field phone, I decided to call the sarge, who had gotten an early out and moved from Oregon to Southern California. He had been a consistent letter writer and became my constant companion for that week, helping me get used to "The World" again—getting my hair cut, shopping, surfing, kite flying, attending a live theater production, etc.—the best definition of "sweeping me off my feet," ending by proposing marriage. I wasn't ready to make that type of commitment after only one week of a "normal" dating relationship, but convinced him to try living together for six months. He agreed to find an apartment while I checked out of the Red Cross in DC and visited my family in West Virginia, returning to California in a month.

For about a month before leaving Vietnam, I had a recurring bad dream of being in a social setting and finding myself between a veteran and a war protester/peacenik. On the flight to DC I had to change planes at O'Hare in Illinois. While sitting in the waiting area, a G.I. in uniform with a 1st Cav patch approached me and said that he remembered me from visiting his unit out in the rice paddies. I was floored when he recited the name of a fellow DD, the game we played with them, and where each of us was from. He sat, and we waited together, ignoring curious stares. I felt very protected and at the same time protective of him. We sat together on the plane; I insisted that he get the window. I was in the middle seat, and a young woman who looked like a student was on the aisle. The G.I. and I continued our conversation, comparing places we had been in Vietnam, R&R's, what was next (he was on leave to visit his wife and children in Virginia and prepare them for going on his next assignment, Germany), etc. About a half-hour into the flight, the young woman tapped me on the arm and said, "I couldn't help but overhear you talking, you were in South Vietnam? What were you doing there?" So I explained that I was with the American Red Cross as a recreation aide. Thinking maybe I could do a little recruitment, I briefly described what we did. She seemed stunned, so I asked if she was a college graduate or student. She said she was a student at Georgetown

University, returning from an SDS (Students for a Democratic Society) seminar at USC. Thinking my bad dream was coming true, I fell silent, hoping the G.I. next to me did not know what the SDS was. She then said, "Why in the world would you spend a year entertaining baby killers?" Before I could say a word, the G.I. reached across me and said to her, "She was there to remind us of home, but not about coming home to bitches like you." Well, that ended all conversation. He patted my arm and said, "Don't waste your time trying to explain."

I didn't want to come home unless everyone came with me (especially my brother, John). I felt pretty washed out emotionally when I got to Travis AFB, so the demonstrators there didn't have much effect on me. I dismissed them as being ignoramuses and cowards who had no clue what they were protesting about. Once I decided to spend my "free" week in Malibu with friends, I was excited and happy. Then began a roller coaster ride . . . sad to learn one of my "big brothers" who was an F-4 pilot on TDY was lost on a mission over North Vietnam, exhilarated by the whirlwind romance of the sarge, happy/relieved to be among friends who didn't ask too many questions, numb at the shopping mall with so many ways to spend money, pinching myself while sitting on the beach having flashbacks to the beach at Cam Ranh Bay, wondering if I would wake up there, excited about the freedom to express my feelings.

Treatment by the public was a mixed bag. I was never spit on as the guys were. I already mentioned the protesters at Travis and rudeness of the SDS student on the plane. I wasn't prepared for being ignored and avoided by others while in my Class A uniform. Once in Malibu, dressed in civilian clothes at a brat cookout, a person I didn't know called me a camp follower after hearing where I had been for the past year in a very derogatory way, and was ejected from the cookout by the host. I have since learned from fellow DD's they got the same type of derogatory comments. We still get them at times from people who don't know the strict rules we lived under while in-country.

I had to check out through National Headquarters of the American Red Cross in DC. I spent two days there while they tried to get me to accept a position at Valley Forge VA as a hospital recreational aide, but

I was just too drained emotionally. I had also made a commitment to my sarge to be back in California in a month. My immediate family had moved from Ohio to West Virginia during the year I was away. It was a quiet homecoming with just my grandparents, parents, and younger brothers, with a lot of hugs. The only questions raised were if I had got to see my brother John before I left. He had received a three-day in-country R&R to see me at Cam Ranh Bay between Christmas and New Year's about a month before I left Vietnam, and I had called him when I got to Saigon right before catching the Freedom Bird home.

Reconnecting with old friends was a mixture: one (who had been classified 4F) said that my going to Vietnam inspired him to accept a teaching position for three one-room school houses in the hollows of Kentucky; another who was a teaching assistant at Kent State said that my experiences were my sure path to Hell; a childhood friend who enlisted in the Navy and served in the Mediterranean wanted to know everything about my experiences; a sorority sister said she wouldn't have the "guts" to admit to what I had done. Two pre-Vietnam boyfriends had different perspectives: one had won a sales contest and saved the prize, a trip to the Bahamas, to take me to "decompress" when I got home. Needless to say, I declined. The other invited me to join him at the commune where he was living. My grandparents treated me with respect and love. My parents treated me like I left my brains in Vietnam. My father even suggested that I had "shell shock" or "gotten into" drugs while over there, saying he had read all about that in Time Magazine and seen news programs talking about it.

Home

I was definitely disconnected from my parents to the point that my father "disinherited" me when my sarge and I invited them to our California wedding that June. The only immediate family member attending our wedding was an aunt. Brother John came to visit on his way home, and I suspect tried to soften Mom and Dad's stance about me, without success. (Although, Mom did send us a wedding present.) My grandparents expressed their love and support. That November my husband and I appeared on the "Newlywed Show" and ended up winning (two motorcycles). I let brother John know of the scheduled airing date, and he passed it on to our parents. Number three brother was graduating from high school the following June, and we were invited home by my mom for that, and celebrated our first anniversary. The parental disconnect was over after a year-and-a-half, and they welcomed my husband into the family.

When I returned east to take care of family in 1993, I found that most friends were still my friends and had even forgotten that I had ever been in Vietnam. I rarely talked about it even with family (mainly brother John). I guess you could say that I disconnected from that year and the experience. John says that I put a cork on that memory bank. That changed in 2007 when I received a phone call from a Red Cross classmate, inviting me to the premier of a Donut Dollie documentary, "A Touch of Home," at Arlington National Cemetery's Women's Memorial over Veterans Day weekend. She had used the wonders of the internet to locate me, saying that I had been Missing in Action for many years, while fellow donut dollies had searched for me. I was able to attend the premier and reconnect with two other classmates, and that cork jiggled loose. In the past eight years, I have gradually reconnected and connected with some of the most amazing women. Sadly, time is taking its toll on our numbers. That is one reason I agreed to speak publicly about my Vietnam experiences for the first time this past spring at the Ohio DAR State Convention. A former college friend is the Ohio State Regent and remembered that I had served with the Red Cross in Vietnam. I was their keynote speaker and received a standing ovation. I was overwhelmed.

My husband and I were living in Orange County, California, in 1975 when Saigon fell. We didn't talk much about Vietnam, I think because we knew what the other had been through. We watched the TV news together that night with disbelief. How could that happen? We were both angry and felt that all the deaths and sacrifices were for naught. I worried about the Vietnamese that had helped us (specifically three papa-sans at the Red Cross Recreational Center at Freedom Hill, Da Nang), and our young Chinese family neighbors in Da Nang. They had shared their story of escaping to Hanoi when the communists took over China, then fleeing Hanoi to Da Nang when the communists took over North Vietnam. When the desperate efforts of so many trying to escape were televised, I prayed that they were among those who made it out.

I don't think I have had any major problems due to my Vietnam experience. A couple of minor things like a spot where I had jungle rot (diagnosed by a Marine doc. while still in-country) left a persistent dry patch of skin to this day. I also have had a bad dream that started the first year I was back in The World, at least once a year and some years several times, of something that never happened. I had some scary moments while there but nothing like this bad dream. I am at Cam Ranh Bay Army, probably one of the safest posts there. We lived in three trailers in the officers' compound. It's the middle of the night, and we get incoming. There is no bunker, so the three of us in my trailer crawl underneath the trailer, and the trailer gets a direct hit and is burning all around us. Then I wake up. Nothing like this ever happened. In fact, our area at the base never got incoming while I was there. The older I get, the less frequently this dream happens.

Looking back

About the war: the mission was righteous, the execution was hampered by politics (whoever heard of a "no fire" zone in a war, especially a guerilla war?). I don't understand why the United States did not include a military presence a la South Korea in the peace negotiations, nor why we did not live up to the military terms of the peace treaty. We were winning the war militarily, but you wouldn't know it according to the media. I think some lessons were learned, but worry that there is still too much politics involved in recent military decisions.

I personally think we should go back to a two-year universal draft for both men and women, similar to Israel's.

I would do it again, given the opportunity. It was a life-changing experience in much the same way as it was for the men. In some ways, the experience forged a stronger me. I worked my way up the corporate ladder to become the first female officer of a large insurance company, which I attribute to being in a war zone for a year. I had no tolerance for bullshit and was known for shooting straight from the shoulder.

I have come to treasure the time I get to spend with fellow Donut Dollies and Vietnam veterans. I never thought what I did was a big deal until Veterans Day weekend, 2013. It was also the 25th anniversary of the Women's Vietnam Memorial and my first visit to the Wall. The Donut Dollies had a reunion that weekend. I was overwhelmed by the support, respect, affection, and appreciation expressed by veterans. The entire costs of transportation (shuttles to the various ceremonies), and banquets held were underwritten by various veteran groups. By the time I got home, my shoulders were bruised from all the hugs received.

I considered myself sort of a tomboy, being the oldest and having four younger brothers: I had to teach them a lot. I took the basics (like hot, clean water) of our American life for granted. I was raised a Christian but was a child of the '60s, questioning that faith. I had no concept or knowledge of real danger or combat. My Vietnam experience changed all those things. I am so thankful for the protection and respect shown to me as a lady by the majority of men serving there. I was probably a prime candidate for the Women's Movement, who were burning their bras while I was over there, but no way would I become one of them after my Vietnam experiences. When you are sure that you will die within seconds or moments, you define your religious beliefs, and they are with you forever when you survive. I had no opportunity before Vietnam to know what compassion was. It taught me compassion for all and real appreciation for all the blessings I have been given. The most important one is to live in freedom.

I am so grateful for the excellent protection our forces and allies gave to us while in-country.

I felt like I'd failed, was scared, and I had no idea what I was going home to.

Rene' Johnson

American Red Cross Donut Dollie

I'd always wanted to be a nurse, never anything else. When I started college, I found out about the Army Nurse Program that would pay my way through college, and thought that I would sign up at the appropriate time. However, due to circumstances beyond my control, I found that I would not be a nurse. So, I drifted on through the social work program. By my senior year, I had so much conflicting input from so many sources—my father, men I knew at Fort Benning who had spent one,

maybe two, tours in Vietnam; professors, student demonstrators, the press—I decided that I had to go to Vietnam to find out why we were there, what was really going on. So, I took the AFOQT (Air Force Officer Qualifying Test) with all the seniors on the FSU football team who didn't want to be drafted! (Guess who was the only person in that room to pass?) I selected Intelligence, got it, and had a reporting date for Randolph AFB the following April. Then, a friend of mine who was already in told me that, as a woman, I would never go to Vietnam. I'd be in CONUS, maybe Europe, doing photo analysis. What?

So, I was bitching and complaining to one of my study partners in a class, who just happened to be on educational sabbatical from the Southeastern Area Red Cross office in Atlanta. He told me about the Supplemental Recreational Activities Overseas program (SRAO, better known as "Donut Dollies"). The SRAO program required that we be college graduates. I was flown to Atlanta, ostensibly for an interview, but was pretty much hired on the spot, and I signed all the paperwork within an hour of my arrival. I called my Air Force recruiter and told him that I'd changed my mind and why. He couldn't offer a better option.

I was very excited when I found out that I would be going to Vietnam! That is why I joined, and soon I would be on my way.

I arrived in Vietnam in April 1969. I was first assigned to the Americal Division at Chu Lai where I remained until October. I then was assigned to the 25th Infantry Division at Cu Chi. From Cu Chi our unit also visited the last brigades of the 9th Infantry Division and the 1st Cavalry Division from October 1969 to April 1970, when I came home.

In September 1970, I returned to Vietnam with the Army Special Services, and was assigned to the Service Club at Bear Cat until November 1970. Then I was at Camp Baxter, Da Nang, from November 1970 to February 1971. From February 1971 to July 1971 I was at Phu Loi, at which time I returned home.

Coming Home

The first time I came home was fine. I came home with two other women, chatting and expressing how much we would miss "our guys." Our first stop was Hawaii. One of the "gun bunnies" (artillerymen on the big guns, 105s, 155s, etc.) with the Americal Division had given me his phone number in Hawaii and told me to call if I was ever there. The other Donut Dollies and I had decided to stop for a few days to decompress, so I called him when we arrived in Hawaii. He and his brothers showed us the Hawaiian side of Oahu, not what we would have seen as tourists. But we sat in front of the TV for hours when they weren't showing us around, just to watch the commercials!

I then flew on a commercial flight to San Francisco. I did not have any negative experiences like the guys experienced: no spitting, no catcalls. I spent a few days in San Francisco, bewildered by what I was seeing of the anti-war demonstrations, and confused because of the lack of support for our men who had just gone into Cambodia. I was very frustrated because I wasn't there for "my guys," the men of the 25th Infantry Division. I went home, looked into jobs, and ended up staying with the Red Cross, assigned to Camp Lejeune. The Marine mentality was something that this Army brat had never experienced, and when offered a job back in Vietnam, I accepted immediately.

The second return was horrible. When I returned the second time, I flew into Travis with the guys, but in civilian clothes, so was not harassed. It was unreal. I was depressed, probably suffering from PTSD, in a bad place mentally. I felt like I'd failed, I was scared, and I had no idea what I was going home to. I'd discovered in the few months between tours that I did not relate at all to women or non-veterans of my age group. I knew that I would be alone.

A USAID friend from Laos flew up to San Francisco to meet me and tried to give me some time and support to sort things out. But I was too far gone at that point and made him go away.

I didn't interact with "the public." And I can't remember a whole lot about coming home the second time after I left San Francisco. I called my parents from the San Francisco airport just before getting on the plane to head east. I can't remember the trip home or any details of my arrival.

My parents were in the process of moving from Georgia back to the home where they planned to retire. I didn't "come home," I went to where my parents were, where I didn't know anyone other than a few relatives.

I was not comfortable at home. Not then, not now. My family moved to a very small town in Florida when I was 16, where I graduated high school in a class of 23. I was an outsider then, having moved there for just a year while my father was in Vietnam, and now, six years later, here I was, a woman who had voluntarily gone to Vietnam—twice!—with the same small-town people and attitudes surrounding me. Most were never close friends. The ones that I was close to had moved away, but the ones remaining were polite, at best. Because I was already exhibiting symptoms of PTSD (which hadn't even been identified yet) I couldn't go into a crowded room, and over-reacted to any sharp sound, etc. I was, in their minds, and even my own, weird. My parents thought that I was using drugs (I wasn't), my aunts and grandmother tried just to treat me like they always had, no questions. They were the best.

Home

Looking back, my wartime experiences changed me immeasurably: politically, socially, personally. I felt disconnected with my family and friends, and that has never ended.

I remember April 1975, when Saigon fell. I was working in Thailand, at Camp Samae San, near U-Tapao. We were getting so much information daily. Some members of the Joint Casualty Resolution Center had wives in Vietnam that they were trying to get out. The babies and little ones were staged through U-Tapao on Operation Babylift, and I helped some with their care while they were waiting to move on. It was all anyone could talk about for days, maybe longer, as the NVA moved south. It was devastating! All those years, all those lives lost: friends, faces that I knew without knowing their names, untold more with life-affecting injuries . . . totally wasted!

My problems began in Vietnam before the end of my second tour. They've lessened, altered over the years, but, knowing what I know now, I was exhibiting classic PTSD symptoms. But no one knew, no one recognized that my issues were not totally of my own doing. I was/am considered weird, difficult, any number of adjectives.

I sought help, but until 2012 I never found anyone who recognized what was going on, or knew what could or would help. Diagnoses included bipolar disorder, clinical depression, others. Meds didn't help. Finally, understanding the problem, learning to recognize symptoms as they appeared and trying to redirect, move on forward, did make a positive change in my life. Forty-four years later. Forty-four lost, wasted years.

As far as Agent Orange issues, I have peripheral neuropathy in my feet, and type 2 diabetes (no one else in my family, either side, has ever had diabetes.)

I was never "for" war, but now I am totally against it! My first year was a year of almost constant highs, adrenalin rushes, good times working with the guys. Some down times also, but I felt like I was doing something that had meaning. The second year, the drug use in Vietnam had changed radically from pot to heroin/opium, much, much more of the latter, and a total change in the attitudes of the guys. The anti-war movement back home, the stories of how guys were treated upon their return, the disallowing of action against the VC/NVA . . . morale was awful, and I felt it. I couldn't do anything about it and was affected by it. I became privy to some information on drug sources, and my life was threatened. I left two months early, pretty much a basket case.

If I had it to do all over again, I'd go back to that first year in a heartbeat. I wouldn't even consider the second year.

Chapter 15 - Final Thoughts

I remember my last day in Vietnam. It was late May 1972, and I was stationed at Vung Tau after having spent 17 months at Phu Loi. I had been there for almost two months. Due to a "misunderstanding" with my commanding officer I was detailed to ride shotgun on convoys to Long Binh during the day and perimeter guard at night. The week before I left I was relieved of my convoy duties and cleared post, but was still required to do perimeter guard. I was out on the bunker line the night before I left Vietnam.

My position was a one-man foxhole. I had my rifle, a field phone, and two flares. The next position was about 50 yards away. Due to a lack of manpower this late in the war, the perimeter was moved back. We didn't have enough men to properly guard the base, but we did what we could. I took my position at around midnight and relieved the evening guard. I was sick, suffering from amoebic dysentery. I was tired in my body and my soul. And, I was angry. I wasn't angry at anything in specific, I was just angry.

I sat in my hole and watched the darkness. Watched and waited. Waited for the unknown. In the distance, I could see a firefight in progress. I could not hear it but I could see the exchange of orange and green tracers. Tomorrow, I would be gone but the killing would continue. No end in sight and no way to achieve any kind of victory. It was obvious that no one cared anymore. No one in the military wanted to be the last one to die in Vietnam and no one at home wanted anything to do with the war or us.

My last night—I thought I would be scared, nervous, apprehensive, elated, or happy. But, in reality, I didn't give a shit. I sat in that hole and slowly watched the sun rise over the South China Sea. The dawn of a new day. My last sunrise in Vietnam, the last of so many. I remember thinking about this, and about the enormous waste. All the dead, all the wounded, all the lives changed forever. For what? The sunrise was scarlet and beautiful. The sky was clear and there was a slight breeze coming from the South China Sea. I sat with my back against my foxhole with my rifle across my knees and watched the stunning beauty of the sun slowly rising in the east. What a contradiction. The beauty of nature, and yet, in the distance, the deaths of the men in the firefight I had watched earlier. In truth, the entire war was a contradiction, and still is today.

I walked over to the company area, went to supply and turned in my rifle, helmet, ammo, and flares. I walked over to the airfield and caught a hop to Tan Son Nhut. As I flew over the rice paddies, I looked down and appreciated the beauty of this country so ripped apart by war. I checked in, showed my orders, showered, changed into khakis, and reported for my flight out of there. My war was over.

We young men and women went to war completely clueless as to what we were getting into. We were naïve and fearless, and we believed we were going to do our duty for our country like veterans of the wars before. We remembered the stories of our fathers and uncles coming home, victorious, after having made the world a better place. But in Vietnam we came to realize that we weren't accomplishing anything. It seemed we were just trying to kill more of the enemy than they were killing of us. We would fight to gain ground only to be ordered to abandon it later.

We saw all the carnage, the horror of our wounded buddies, and experienced the pain of their deaths. When the time came to rotate out and our tour was up, we returned alone without counsel or decompression time, thrown back into the real world only to discover not only were we not appreciated, but, rather, we were vilified or ignored.

One or two years of our young lives spent in a foreign land, scared, subjected not only to enemy fire but to dangerous chemicals and moral injuries. Why? More than 58,000 dead! No one cared. No one appreciated it. It left us with a feeling of emptiness, anger, and terrible sadness. Wasn't there something more to this? Time went on, and the VA was unresponsive to the many physical and emotional claims we Vietnam vets suffered for years. Many of the stories in this book illustrate the toll of this war, whether it was divorce, substance abuse, or suicide. I hope that by reading these stories you have a better understanding of what it was like to have been one of these young men and women in harm's way, unappreciated.

Back stateside, the news was filled with campus unrest and protest marches. Some students burned their draft cards, and some fled to Canada or Europe, while others married, attempting to avoid the draft. Others managed to get a student deferment and at least postpone the possibility of military service. All in all, the students lived a pretty comfortable life, while those who served did their country's bidding and went where ordered. The "silent majority" was basically that: silent. They watched the war on the evening news and spoke about it among themselves, but it didn't really affect them, unless they had a relative who was serving. They had no personal investment in the war. The economy was buzzing, and life for the most part was good. When the evening news was off and the day's newspaper thrown away, the war was forgotten.

For a serviceman in that era, Vietnam was all-encompassing. Even stateside its damage could not be ignored. When I started basic training in January 1970, nine drill sergeants out of ten in my company had already served in Vietnam. Four were wounded there. When I broke my foot in basic and was admitted to the orthopedic ward, I was the only one on the ward who hadn't been wounded in Vietnam. These men were all busted up. Universally, they were in good humor while they healed. But so many had other wounds, many of which were invisible.

One soldier in particular comes to mind. He had been wounded by shrapnel, but also suffered psychological injuries. He would lie on his bed or sit and stare into space. He was very quiet. I remember his wife and parents came to visit him. He did not acknowledge them, just stared into space. Then the staff brought lunch. He stared at the tray and started to pick up each item on the tray and identify it. "This is my milk. This is my spoon. This is my soup." His wife and parents started to cry. It was deathly quiet on the ward until they left. No one mentioned what had just happened. Some wounds leave deep scars that are not visible to the naked eye.

Vietnam was part of our lives no matter where we served or what job we did. If you didn't serve in Vietnam, you knew others that did. You were either training for Vietnam, waiting for orders to go there, or riding out your time until discharge. Some were hoping not to go and others hoping they would. It didn't matter if you enlisted or were drafted. It didn't matter if you were Army, Navy, Air Force, Marines, or Coast Guard. It was always there. The war just went on and on: there was no end in sight. For most, it was going on when you entered service and still going on when you were discharged. You weren't in for the duration like World War II. What were we doing? Were we making a difference? What the hell was going on?

We went to Vietnam aboard ship, military transport, or commercial airliner. Innocent, curious, excited, nervous, afraid, we had no idea what to expect. What unit would we be assigned to? What part of the country would we be in? Would we work the MOS we were trained in? Would we perform well under fire? Would we be wounded? Would we be killed? Would we ever see our families again?

We faced the hazards of duty in a war zone. And, believe me, there were a lot of ways to die in Vietnam. Enemy fire, of course, was the thing we feared and expected the most. We didn't think of being killed or maimed by friendly fire or an accidental discharge by a fellow soldier. We didn't think about being killed in a plane or helicopter crash due to bad weather or mechanical failure. We didn't think about vehicle accidents. It didn't occur to us that we might be felled by malaria or other tropical diseases. We never thought that we might be distracted and walk into a tail rotor.

We hunkered down through the firefights, combat assaults, rockets, and mortars. We suffered through the heat, humidity, and monsoons. We missed holidays, birthdays, and anniversaries. We missed the passing of parents, grandparents, and sometimes friends. We dealt with the Dear John letters from wives and girlfriends who could not wait a year for us to return. We dealt with the loneliness and longed for home.

We did our duty and we went when ordered. Some volunteered, but most just went because they received orders to go. We didn't want to go halfway around the world for this war, but we went. It was our duty, our job. We didn't cut and run. We served our 365 days until that beautiful Freedom Bird would fly us out of there. Then we came home. Home.

When we left Vietnam for home we were excited, nervous, apprehensive, happy, and relieved. When we went overseas we had no idea what waited for us, and returning home we felt the same way. The reception given us by our peers was a shock. We were either treated poorly or completely ignored as if nothing had happened. It was a depressing feeling. There was no, "Thank you," or, "Welcome home, glad you made it." There was venom, scorn, or indifference. No one asked what it was like. No one asked what was actually going on over there. It seemed that the public was either against the war and those fighting it, or they didn't care. At times, it seemed like no one even knew there was a war going on. When I came home in 1972, the US involvement in the war was rapidly ending. For those soldiers still in Vietnam, the war was very real. The Vietnamese had taken over almost all the fighting, but there were still Americans dying. I was back in the States and was sent TDY to Aberdeen Proving Grounds in the fall of 1972. Discharge was only months away. I was in a club with a couple fellow soldiers. I am sure we were the only G.I.s there. Everyone was dancing, drinking, and laughing. I was overcome with a sense of depression and sadness. It was like the war wasn't happening, like it never happened at all. I went to a phone booth and called my dad, but I didn't know what to say. I was at a loss for words. It was like it never happened. Over 58,000 Americans died and no one seemed to care. The thought of it was overwhelming. No one gave a damn.

The country wanted to put Vietnam behind it, forget it, move on. The war divided our country, possibly the biggest division since the Civil War. The anti-war movement felt vindicated. I am sure they spoke proudly of their part in ending the war. The hawks shook their heads and blamed the military for losing the war. To the anti-war movement, we were a symbol of the war, and they directed their disdain and hate at us. Even though many of those drafted came from their ranks, they were lumped in with everyone else and treated with disrespect.

Many establishment organizations, like the VFW and American Legion, called us losers and discouraged us from joining their ranks. What the hell happened? Was the country ashamed of us? Why did they send us over there to begin with? We were the casualties of the war. Those who did not participate moved on and continued with their lives, unbruised and undamaged. They finished their education and started their careers while we were away. We who served and survived had to catch up, they were a step ahead.

We faced other issues besides rejection and scorn. Our ranks have been ravaged by Agent Orange-related illnesses. Cancers, heart disease, and lung issues have claimed the lives of men and women alike. No one who served in Vietnam was immune. The legacy of our exposure has been handed down to our children and grandchildren. No one has any idea how many future generations will become casualties of Vietnam.

We returned home, alienated by the society we so wanted to rejoin. We left some of the closest friends we ever had or will ever have. We anticipated seeing our family and our old buddies. But, things rarely work out the way we envision. Our families danced around the time we spent in Vietnam. In most cases, it was hardly mentioned. Our old buddies had moved on or in some cases were still stuck in a high school frame of mind. Some were anti-war and treated us like we were the ones who started the war. Most really didn't give a damn. We were in limbo, thousands of miles away from our friends in Vietnam and figuratively thousands of miles away from "home." We felt completely alienated from everything we knew and used to know. As Jimmy Taylor said in his interview, there was an aloneness about this war, and that aloneness followed us home.

Time has gone on and we veterans are getting old. The effects of war seem to hit hard when we start to approach retirement age. The kids are raised; our career is over or as good as it will probably get. We now have more time to think about past events that changed our lives. The prospect of death once again is becoming a reality. Some are haunted by memories. Moral injuries surface more and more as we age. In war, veterans experience actions that are against their moral and spiritual beliefs. We are raised with the notion that, "Thou shalt not kill," yet in war we do. How do we live with and justify these acts? Peace and forgiveness can only come through spirituality. Some find that peace and others never do.

A new generation of war veterans has come on the scene and is being accorded the respect they deserve. The nation as a whole recognizes them and tells them how their service is appreciated. We are even included in some of the "Thank you for your service" accolades. Vietnam veterans are now coming out of the closet. We are wearing our "Vietnam Veteran" ball caps and jackets. But, the hurt and pain of our homecoming is still a festering sore. We watched when the new war veterans entered the airport terminals greeted by applause and cheers from an appreciative public. We were all happy to see them get the welcome they deserved. But, many of us, deep down, harbored some resentment. It embarrassed us, but it was there.

I was at a golf outing in Myrtle Beach a few years ago. There were 24 men in attendance, all in my age group, and I was the only Vietnam vet. At dinner one night I started telling the ten at my table about this book project. One had served in the National Guard but the rest had no military experience. I told them how I was occasionally thanked for my service but have never been thanked by anyone of my generation. They were shocked. But, when asked, they all admitted that they had never thanked a Vietnam vet for their service. Of course, they then all thanked me for my service. It rang hollow. To this day, I don't recall ever being thanked for my service by someone of my own generation. The scars of Vietnam are thick. Civilian and veteran alike, I guess. I think that most people just want it all to go away, want it forgotten.

I often wonder what happened to those that gave us the finger and called us names. Do they applaud the troops now as they come home? Do they remember how they treated us, how they bruised our souls? Do they even care? I wonder if any of them will read this book. I wonder if they are proud, embarrassed, or just indifferent. I wonder if they even give a damn.

Time has gone on. For the most part, many of us can't remember exactly what names we were called, but we will never forget how they made us feel. I have never talked to any protesters from those days but I would like to have them explain to me why they treated us that way. I would like to sit down with them, explain my life, hear about theirs, and listen while they explain to me why they treated us that way; why they hated us. We weren't murderers. We answered our country's call. They didn't. Sometimes life's decisions are hard.

I hope this book will help people understand what the Vietnam War was like for all the young men and women who fought it and how it changed our lives. We thought we would at least be welcomed when we came home, but that was rarely the case. After seeing all the horrible things war can do, we came home to ignorance or hostility. Suddenly, we were back with our loved ones, but the war followed us and intruded on our home life. Many became alcoholics, or couldn't sustain a relationship, and some people just couldn't handle it anymore.

One of the saddest days I experienced while writing this was when I received a phone call from the wife of one of the veterans I had interviewed. She told me he had committed suicide. Shot himself. A man, in his late 60s, with a family, so affected by a war so long ago that he takes his own life. Even though he is on the eighteenth hole of life, he can't stand to live another day. Some wounds never heal.

AIT Advanced Individual Training, Advanced Infantry Training
ARVN Army of the Republic of Vietnam
AVEL Aviation Electronics Company
AWOL Away Without Official Leave
C&C Command and Control Helicopter. This would be the commander of a military operation who is orbiting above troop insertions or combat operations.
Charlie Viet Cong
Civies Civilian clothes
CO Commanding Officer
CONUS Continental United States - The World
DD214 Certificate of Release or Discharge from Active Duty. Given to all military personnel upon discharge from active service.
DEROS The date you catch the Freedom Bird and fly home.
Drop Early end of your tour in Vietnam or release from active duty.
Early Out Being released from active duty prior to your original discharge date.
EOD Explosive Ordnance Disposal
ETS Expiration, Term Of Service. Date of discharge
FNG Fucking New Guy. A newly arrived trooper.
Freedom Bird Any aircraft (usually a commercial chartered airliner) that takes you out of Vietnam
FSB Fire Support Base
H&I Fire Harassment and Interdiction Fire. Artillery fire directed at possible enemy targets
LZ Landing Zone
MAC Military Airlift Command
MACV Military Assistance Command, Vietnam
MARS Military Affiliate Radio Service. The MARS system offered soldiers and sailors a way to personally communicate with loved ones back home via the use of a "phone-patch" telephone connection over

short-wave radio. Amateur radio operators across the country patched these calls without cost.

MOS Military Occupational Specialty. The job performed in the military. Every job had an assigned MOS code.

Nomex Flame resistant flight suites

NVA North Vietnamese Army

OCS Officer Candidate School

PRC25 Backpack radio used by the infantry as their main source of communication. Commonly called "Prick 25"

Permanent Party Cadre assigned to a duty station on a permanent basis. Not troops who rotated in and out on a temporary basis.

PZ Pickup Zone

RRG Radio Research Group

RRU Radio Research Unit

RTO Radio/Telephone Operator

Ruck Rucksack

RVN Republic of Vietnam

SAC Strategic Air Command

Sea-Tac Seattle-Tacoma Airport

Short Anyone whose tour is approaching an end. Usually less than 30 days until it ends.

Single Digit Midget A trooper with less than 10 days to go on their tour

Slick Troop-carrying helicopter

SP Shore Patrol

TAD Temporary Additional Duty

TO&E Table of Organization and Equipment

USAID United States Agency for International Development

USARV United States Army Republic of Vietnam

VC Viet Cong

WESTPAC Western Pacific

Xin Loi Vietnamese term meaning "excuse me" or "pardon me" but commonly used by American military personnel to say "sorry about that," "too damn bad," or "tough shit."

Jack McCabe was born and raised in Chicago. He joined the army in 1969, two days after his 18th birthday. He was sent to Vietnam less than a year later, in October of 1970. He extended for a second tour and finally came home for good at the end of May 1972, finishing his three-year enlistment at Fort Huachuca, Arizona.

Upon his return from Vietnam, Jack pursued his education using the G.I. Bill, receiving an associate degree in electronics engineering from DeVry Institute. Later, he continued his education by attending night school, and received his bachelor's degree in business and management from Northeastern Illinois University in 1981 at the age of 30.

Jack owned his own business for 20 years and sold real estate for 20 more, before retiring to North Carolina, where he became a certified Peer Support Specialist with a veteran designation. He now works as a volunteer, helping veterans with PTSD, financial challenges, family issues, homelessness, substance abuse, and benefits for veterans.

Jack maintains an active Facebook group for *When We Came Home*, where Vietnam veterans and others share information and advice and make connections. You can find the group here: http://bit.ly/2ujUzDi

66951883R00211

Made in the USA
Lexington, KY
28 August 2017